Fuels and Petrochemicals Division 2019

Held at the 2019 AIChE Annual Meeting

Orlando, Florida, USA
10 - 15 November 2019

ISBN: 978-1-7138-0538-0

Printed from e-media with permission by:

Curran Associates, Inc.
57 Morehouse Lane
Red Hook, NY 12571

Some format issues inherent in the e-media version may also appear in this print version.

Printed with permission by Curran Associates, Inc. (2020)

For permission requests, please contact AIChE
at the address below.

AIChE
120 Wall Street, FL 23
New York, NY 10005-4020

Phone: (800) 242-4363
Fax: (203) 775-5177

www.aiche.org

Additional copies of this publication are available from:

Curran Associates, Inc.
57 Morehouse Lane
Red Hook, NY 12571 USA
Phone: 845-758-0400
Fax: 845-758-2633
Email: curran@proceedings.com
Web: www.proceedings.com

TABLE OF CONTENTS

571591 Demystifying Mass Transfer in Electrolyzers through Rapid Prototyping with 3D Printed Parallel Plate Cells

As a result of the global energy transition, a large percentage of the world's power supply will become electric. For electrochemical engineers this transition generates a sea of opportunities. After all, it is electrolyzers that use electricity to produce some of life's essential chemicals. However, fully embracing these opportunities in a sustainable way is challenging: Future electrolyzers must be safer, cleaner, and far more flexible, while simultaneously increasing productivity and cost-effectiveness. Essentially, they will have to achieve more with less.

One key parameter in the performance of electrolyzers is the rate of mass transfer. Most often, electrochemical reactions are mass transfer limited due to the fast kinetics that depend exponentially on the applied overpotential. Numerous correlations in literature exist to describe the rate of mass transfer in a specific reactor.[1] Though, their predictions can vary up to almost an order of magnitude (see figure 1). The reason for this is the large geometric variety between the reactors that were measured. The design of the inlets in particular is a crucial contributor to the rate of mass transfer. [2] Moreover, only a few lab-scale electrolyzers are commercially available, often with sub-optimal designs, which limits further research. As a consequence, it is difficult to establish a generalized description of the mass transfer performance of parallel plate electrolyzers. Furthermore, most existing correlations only describe the average rates of mass transfer and therefore do not consider any local variation despite the importance thereof.[3]

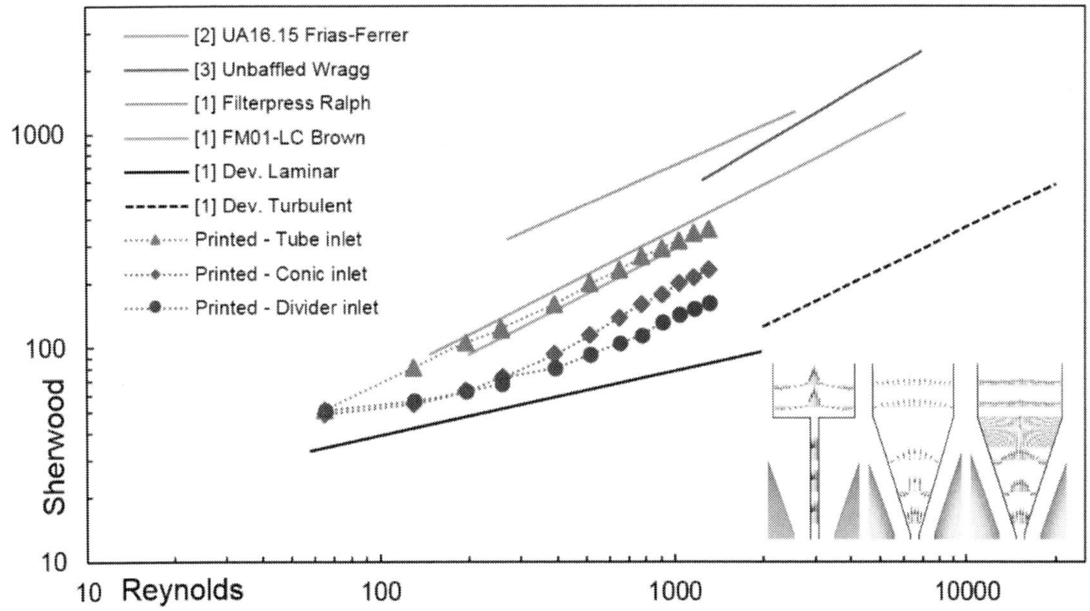

Figure 1: Mass transfer performance in terms of Sherwood vs. Reynolds. Up to an order of magnitude of difference in performance is visible between reactors. The red triangle, square and circle markers denote a selection of inlet manifolds tested with the 3D printed prototype. Small insert: sketches of the inlet manifolds, from left to right: tube inlet, conic inlet, divider-type inlet.

The goal of our study has been to demystify the fundamentals of mass transfer in parallel plate electrolyzers. As this would have required testing a large geometric variety of cells, traditional manufacturing techniques were impractical due to their lead time and cost. Therefore, we have used 3D printing to rapidly construct many different reactor prototypes. Moreover, each prototype could be finetuned to the exact specifications each measurement required. This way, each geometric parameter could be varied and investigated individually.

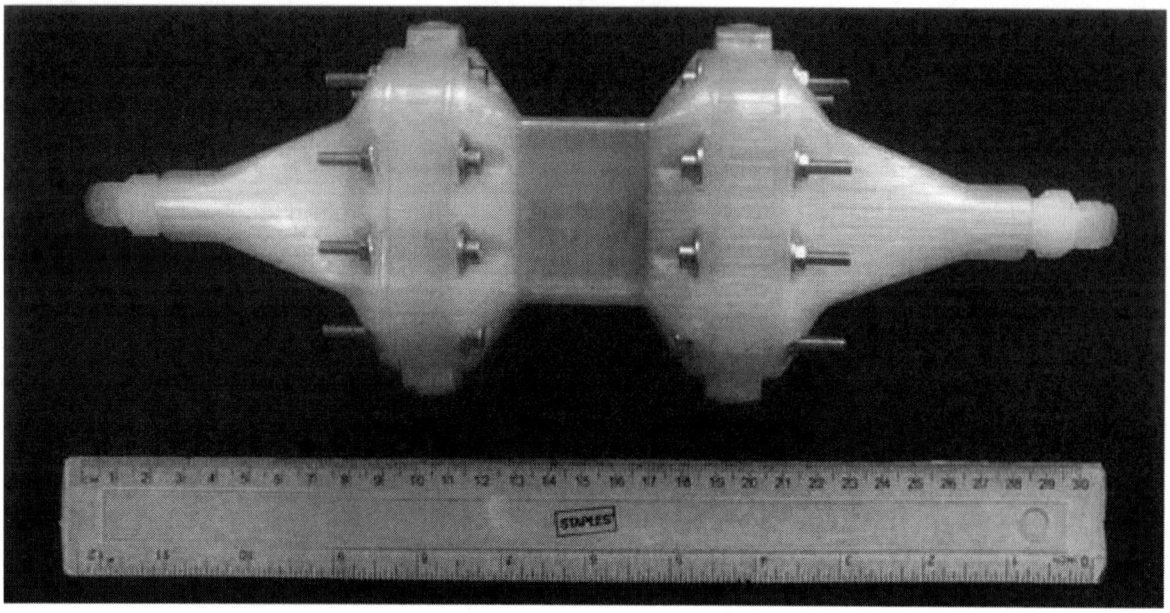

Figure 2: 3D printed electrolyzer prototype for mass transfer studies. The electrode area is 40 cm^2.

Nearly the entire reactor can be 3D printed in less than a workweek's time (see figure 2). Inlet manifolds, the channel and the rubber sealings could all be printed using commercially available equipment. Furthermore, by using 32 independently operated electrode segments we were able to measure the mass transfer performance on a local scale. Sherwood – Reynolds correlations were established for designs with different shapes of turbulence promoters, 3D electrodes and inlet manifolds. The effect of an increased or decreased hydrodynamic entrance length was also investigated.

In the end these prototypes lead to enhanced understanding of electrochemical reactors, allowing us to intensify processes such as alkaline water electrolysis. As a proof of concept, we are developing a 3D printed lab-scale electrolyzer for alkaline water splitting which can be used to test the effect of geometry on the production of hydrogen directly.

References

[1] T.R. Ralph, M.L. Hitchman, J.P. Millington, F.C. Walsh, *Mass Transport in an electrochemical laboratory filterpress reactor and its enhancement by turbulence promoters,* Electrochem. Acta, **1994,** pp. 591-603.

[2] A. Frías-Ferrer, J. Gonzálex-García, V.Sáez, C. Ponce de León, F.C. Walsh, *The effects of manifold flow on mass transport in electrochemical filter-press reactors,* AIChE J., **2008,** pp. 811-823.

[3] A.A. Wragg, A.A. Leontaritis, *Local mass transfer and current distribution in baffled and unbaffled parallel plate electrochemical reactors,* Chem. Eng. J., **1997,** pp. 1-10.

574364 Additive Manufacturing of Catalyst Support Structures with in Operando Adjustable Mass Transport and Flow Characteristics

In conventional randomly packed bed reactors, the flow and mass transport characteristics are predefined by the particle diameter and the tube diameter. Structured reactors offer the opportunity to tailor mass transport and flow field according to the needs of the reaction system. In this regard, additively manufactured so-called Periodic Open Cellular Structures (POCS) as catalyst supports have shown their advantages over conventional randomly packed bed reactors in terms of heat management and pressure drop [1, 2]. However, since the internal flow field, pressure drop and therefore the resulting mass transport characteristics are mainly influenced by the geometrical properties of the unit cell, for POCS the resulting transport characteristics of the structure is already determined and fixed by the choice and design of the unit cell. To achieve more flexibility, we have recently proposed a new type of so-called interpenetrating POCS (interPOCS) which are formed by inserting a diamond unit cell based structure in the cavities of a second diamond cell structure. These interPOCS allow for in operando flow field and mass transport adjustments by shifting the (relative) position of the loose structure [3].

With detailed computational fluid dynamics (CFD) based simulations with an additional in-house particle-tracking algorithm implementation we systematically investigated the flow field and mass transport characteristics within the interPOCS in dependency of the relative positioning of the loose structure. The highly adjustable structure allows for a broad variation of mass transport characteristics in a catalytic reactor without changing the system's periphery. This functionality enables mass transport optimization for a given system and represents a completely new application of additively manufactured catalyst support structures for highly flexible in operando tuning of flow field and mass transport characteristics.

References

[1] C. Busse, H. Freund, and W. Schwieger (2018). *Chem. Eng. Process.* 124, 199-214.

[2] A. Inayat, M. Klumpp, M. Lämmermann, H. Freund and W. Schwieger (2016). *Chem. Eng. J.* 287, 704-719.

[3] G. Do, T. Stiegler, M. Fiegl, L. Adler, C. Körner, A. Bösmann, H. Freund, W. Schwieger and P. Wasserscheid (2017). *Ind. Eng. Chem. Res.* 45, 13402-13410.

561588 Role of 3D Printed and LED-Driven Photostation in Photocatalytic and Photo-Fenton Activity of Iron Oxide Doped Graphitic Carbon Nitride

The degree of photocatalysis or any photolysis reaction depends on the intensity of light. In most cases, a major portion of light is not used as it does not reach the reaction zone thereby wasting a significant portion of light and energy, as well. In order to handle this situation, we have designed and 3D printed a photostation with complex parabolic reflective surface, so that all the incident light beams can reach the reaction zone. The photostation is fitted with two chips-on-board (COB) light emitting diodes (LEDs) reaching the total (visible) light intensity within the reaction zone over 290,000 lux (For comparison, it is commonly reported that the sun emits anywhere from 32,000 to 100,000 lux onto the Earth). Light simulation analysis confirmed that all incident light beams can reach the reaction zone of photostation. In order to test the performance of this photostation, we have used 0.5 to 2 wt.% iron oxide doped graphitic carbon nitride (g-C_3N_4) to degrade a model water pollutant, dicamba (3,6-Dichloro-2-methoxybenzoic acid), which is common herbicide, in presence of traces of hydrogen peroxide and visible light. It has been confirmed experimentally that the rate of degradation of dicamba is directly proportional to the light intensity and geometry of photostation thereby confirming the positive role of photostation design. The rapid degradation of dicamba under this study can be attributed to the four distinct phenomena of photocatalytic activity of carbon nitride, quenching of electron/hole pairs and generation of additional reactive hydroxyl radicals by hydrogen peroxide, Fenton and photo-Fenton activity of iron oxide component of carbon nitride in presence of hydrogen peroxide and photocatalytic activity of iron oxide alone in conjuncture with carbon nitride. Liquid Chromatography Mass spectrometry (LCMS) analysis of the degradation products showed loss of chlorine from the aromatic ring and evidence of free radical addition reactions in the course of photocatalysis.

557053 Solution-Based 3D Printing of Hierarchical Microporous Polymers for Energy-Efficient Adsorption

Current 3D printing methods have strong limitations in the classes of polymers that can be printed. This means that many polymers of technological interest cannot currently be 3D printed. Here, we demonstrate a generalizable method for 3D printing of viscoelastic polymer solutions and apply it to both porous (a polymer of intrinsic microporosity, PIM-1) and non-porous (e.g., cellulose acetate, Matrimid) polymers. Successful ternary ink formulations that result in the creation of three-dimensional structures with hierarchical porosity require balancing of solution thermodynamics (phase separation), mass transfer (solvent evaporation), and rheology. To demonstrate the utility of this approach, a microporous polymer (PIM-1) that cannot be printed with current 3D printing technologies is 3D printed into a high-efficiency respirator exhibiting hierarchical porosity ranging from sub-nanometer to millimeter pores. These short mass transfer contactors (1.27 cm) were challenged with a flow of toluene vapor (10000 ppm) in N_2 gas. Owing to the precisely designed gas distribution channels, the 3D printed respirator were found to create a pure (< 1 ppm toluene) N_2 stream for 1.7 hours, 6 times longer than traditional structures fabricated from the same PIM-1 material, and more than 4,000 times the residence time of gas in the bed. This solution-based additive manufacturing approach significantly extends the material spectrum of 3D printing technologies. By 3D printing adsorptive materials into precisely designed mass transfer contactor, the performance, energy efficiency, and material efficiency can be significantly improved.

559790 Titanium/Titania Electrocatalyst for Hydrogen Evolution Synthesized Via a Combination of 3D Printing and Anodization

A titanium core/titania coating, hierarchically structured catalyst was fabricated via combining 3D printing and in-situ anodization. 3D printing enabled control of the macro-porosity of the titanium core. Anodization was used in order to grow highly-ordered TiO_2 nanotubes (NTs) on the metal core. The dimensions of the TiO_2 nanotubes were controlled via tuning of anodization parameters. The combination of both fabrication techniques allows for hierarchical control of the catalyst structure.

The performance of the reported 3D Ti/TiO_2 electrocatalyst was tested for hydrogen-evolution reaction (HER) via water electrolysis. Titanium dioxide is a known photo-electro-catalysts for water splitting, but competitive electrocatalytic activity for HER in the absence of light has not been reported. Onset potentials for HER as low as -45 mV vs. RHE were obtained for our electrodes in acid medium, making them competitive with current state-of-the-art electrodes aiming to replace platinum. Tafel slopes between 40 and 52 mV dec^{-1} point towards a Volmer-Heyrovsky mechanism, with the Heyrovsky reaction as possible controlling step.

The proposed electrocatalysts feature the following advantages: (1) the fabrication method for the 3D Ti/TiO_2 catalyst is simple, scalable and can be readily incorporated as printed reactor components into different technologies; (2) the 3D Ti/TiO_2 catalyst features a flow-through architecture that not only enables more efficient utilization of the surface area of the catalyst but that could enable large-scale, continuous-flow hydrogen-production reactor technologies; and (3) minimization of electrical resistance between active surface area (TiO_2 NTs) and current collector.

556480 Advanced Manufacturing Enabling Process Intensification for Catalytic Processes

Advanced manufacturing is a transformative and disruptive approach to industrial production that enables the rapid synthesis of virtually any geometry composed of a wide range of materials, such as polymers, ceramics, and metals. Process intensification is a set of novel approaches to process engineering and equipment design that leads to substantially smaller, more selective, and energy-efficient processing plants. The integration of advanced manufacturing with process intensification accelerates process development and deployment through rapid prototyping and enables new hardware that overcomes transport limitations. Advanced manufacturing (ex. 3D printing or additive manufacturing) is the physical backbone of Industry 4.0 that can take advantage of the conclusions of big data and machine learning by allowing design freedom. This expansion of the material design window can be utilized to reduce GHG intensity for novel or existing processes.

Numerous conversion technologies in the energy industry suffer from rapid coke build-up and subsequent deactivation of a catalytic material that is exposed to a carburizing environment. The time scale of deactivation often determines the reactor technology and process configuration employed. Commercial examples of deactivating catalysts range from seconds in a fluid bed catalytic cracking unit to minutes for moving bed propane dehydrogenation and days for continuous catalytic reformers or even years for steam methane reformers. A common step for these processes is the regeneration of the catalytic material via combustion of the coke. However, the cyclic process of coking in a carburizing environment followed by combustion in an oxidizing environment can often be detrimental to catalyst performance.

This work will discuss how advanced manufacturing methods can enable construction of materials capable of rapidly switching between two different environments with enhanced catalytic activity. Such materials capable of withstanding cyclic carburizing and oxidizing steps can be efficiently integrated into novel reactor concepts. The conversion of light gas is an excellent probe reaction for elucidating this behavior. In an effort to search for robust materials, honeycomb structures composed of metal were fabricated by direct metal laser sintering (DMLS), a 3D printing technique that involves the layer-wise successive deposition of micron-sized powder followed by laser induced sintering. The cyclic operation that alternates between carburizing and oxidizing environments leads to compositional modification at the surface of the monolith channels that imparts in-situ catalytic activity.

575756 Oxidative Coupling Methane Enhanced By Distribution of Oxygen through Porous Medium

Recently, technologies for converting methane into more valuable chemicals are fast emerging due to a large amount of global natural gas reserve along with rapid depletion of fossil fuel. Among the technologies, Oxidative Coupling of Methane (OCM) is attractive since it produces valuable ethylene, a precursor to many industrially important chemicals. One of the major problems in OCM is the decrease of C_{2+} selectivity with methane conversion because CH_4 and O_2 are thermodynamically favored to proceed to full oxidative conversion to form CO or CO_2. In addition, the temperature rises by the exothermic reaction further result in the high driving force to full oxidation. In this context, it is important to develop new processes to minimize deep oxidation of CH_4 while improving thermal management. To confirm the performance of OCM in a membrane reactor, in the present work, a process where methane flows over the catalyst while oxygen is distributed over the catalyst bed through a porous filter was investigated. It was found that both C_{2+} selectivity and yield were improved by distributing oxygen through the filter, as compared to those for a packed bed reactor. Finally, operating conditions including temperature, feed ratio, and flow rate were optimized to maximize C_{2+} yield.

576816 Direct Conversion of Methane to Aromatics Using Metal Carbides

Due to the recent shifts in feedstock, the global petrochemical industry has experienced a significant disruption in the supply and demand of key chemical building. Supplies of ethylene have surged as US shale gas production has boomed, meanwhile the supplies of benzene and other light olefins have declined. Therefore, finding direct routes to convert methane into valuable, and preferentially, liquid products is highly desired. Up to now, various catalytic routes including dehydroaromatization,[1-3]and oxidative coupling to produce ethylene[4] have been investigated but still cannot compete with the well stablished, although highly energy intensive, steam reforming of methane. This indirect route convert methane into *syngas* to then produce most of the chemicals we use nowadays, while the direct conversion of methane is potentially more environment friendly and economical but the challenge lies in stabilizing the catalyst against coking. Metal containing zeolites, especially Mo/H-ZSM-5, have shown promising results for MDA reaction. Commercial prospects for MDA process are hampered by the copious amount of coke deposition on Mo/H-ZSM-5 catalyst at the high reaction temperatures. We believe that using identifying the optimal conditions during the preparation, carburization and testing of the catalyst, along with characterizing the Mo species formed and/or transformed during each step prior and during the reaction might be an effective alternative towards attaining stability, selectivity and activity under higher temperature conditions. As many of the important questions related to nature of active sites, catalyst deactivation mode etc. are still unanswered thus limiting the commercialization of MDA process.

In this presentation, we will present our results related to different pre-carburization methodologies to tune the Mo/H-ZSM-5 catalyst properties with enhanced activity and stability. We control the nature and the amount of carbide species formed during the initial induction period and relate that with the catalysts activity. Stress in the presentation will be focused towards the present Mo/H-ZSM-5 catalyst and the newly emerging family of two dimensional (2D) layered carbides, with an aim to stabilize the existing catalyst and to further explore different catalytic active material for the MDA process.

References

1. Gao, J.; Zheng, Y.; Jehng, J.-M.; Tang, Y.; Wachs, I. E.; Podkolzin, S. G., Identification of molybdenum oxide nanostructures on zeolites for natural gas conversion. *Science* **2015,** *348* (6235), 686-690.
2. Lezcano-González, I.; Oord, R.; Rovezzi, M.; Glatzel, P.; Botchway, S. W.; Weckhuysen, B. M.; Beale, A. M., Molybdenum Speciation and its Impact on Catalytic Activity during Methane Dehydroaromatization in Zeolite ZSM-5 as Revealed by Operando X-Ray Methods. *Angewandte Chemie International Edition* **2016,** *55* (17), 5215-5219.
3. Kosinov, N.; Wijpkema, A. S.; Uslamin, E.; Rohling, R.; Coumans, F. J.; Mezari, B.; Parastaev, A.; Poryvaev, A. S.; Fedin, M. V.; Pidko, E. A., Confined Carbon Mediating Dehydroaromatization of Methane over Mo/ZSM-5. *Angewandte Chemie International Edition* **2018,** *57* (4), 1016-1020.
4. Bhatia, S.; Thien, C. Y.; Mohamed, A. R., Oxidative coupling of methane (OCM) in a catalytic membrane reactor and comparison of its performance with other catalytic reactors. *Chemical engineering journal* **2009,** *148* (2-3), 525-532.

565490 Fe_2O_3@SBA-15 As an Efficient and Stable Catalytic Oxygen Carriers for Chemical Looping Partial Oxidation of Methane to Generate Syngas

Chemical looping partially oxidation of methane is one the most promising techniques to generate syngas. Developing high performance oxygen carriers are the key to this process. In this research, Fe2O3 nanoparticles embedded in mesoporous support SBA-15 (Fe2O3@SBA-15) was synthesized and applied to chemical looping partial oxidation to generate syngas. For comparison, bulk Fe2O3 was also synthesized and characterized. Techniques including XRD, BET, TEM, SEM, and TGA were used to characterize the material before and after reaction. Results showed that Fe2O3@SBA-15 have enhanced stability and reactivity, with a conversion rate 70% higher than bulk Fe2O3. This can be attributed to tube structure of the support SBA-15, which provides large space and surface area for Fe2O3 nanoparticles and prevent the agglomeration of nanoparticles during reaction. In addition, selectivity was also test in TGA. Results showed that very low concentration of CO2 was observed during reaction, indicating high syngas selectivity >99%. This designed material has the potential to generate pure syngas at relatively low temperature. The atomistic thermodynamics methods and density functional theory calculations were carried out to investigate the selectivity enhancement mechanism. It was found activation energy barrier of CHx radicals strongly depends on the particle size, and SBA-15 support has an influence on the surface termination geometry and oxygen vacancy formation, which significantly influences the correlation between the C-H bond activation and the CHx radical adsorption. This study provides guidance on the design of novel oxygen carriers for chemical looping partial oxidation.

578427 Stability of Supported Metal Oxide Catalysts for the Non-Oxidative Dehydrogenation of Propane: An *in Situ/Operando* Raman Study

In this study, we investigate the activity, selectivity, and stability of VO_x/Al_2O_3 and $(VO_x)_m$-$(M_2O_y)_n/Al_2O_3$ (M_2 = Mn, Mo, Zr, Ta) catalysts in the non-oxidative dehydrogenation of propane. We will highlight the promoting effect of M_2O_y for tuning the type and amount of carbon produced during the catalyst's deactivation process. Among the current on-purpose processes to produce light olefins, either platinum or chromium supported catalyst are the most used primarily because of their significant activity and selectivity [1]. However, both the high price and environmental restrictions demand the discovery of alternative catalytic materials to produce light olefins [2]. Both vanadium and molybdenum oxide supported catalysts have already shown promising results in the dehydrogenation of light olefins, [1, 3, 4] although improvements are still needed. The main challenge to overcome in the non-oxidative dehydrogenation of light hydrocarbons is the catalyst's stability during both the reaction and the regeneration process. The chemical nature of the carbon formed during the catalyst's deactivation is directly related to the thermal stability of such carbonaceous species and, therefore, to the regeneration process [5]. *In situ/operando* Raman spectroscopy is a suitable technique to identify and monitor changes of the type of carbon produced during catalytic processes [5]. Our preliminary results indicate that basic metals oxides, such as manganese, promotes the formation of principally amorphous carbon while acidic promoters lead to more graphitic carbon. By monitoring the combustion products (CO and CO_2) using *operando* Raman spectroscopy and Raman-spectrokinetics [6], we aim to gain further insights both in the thermal stability and the chemical nature of the carbonaceous species either deposited on the catalyst surface or reacting with the catalytic component(s) forming, for instance, metal carbides.

References

[1] Liu G, Zhao Z-J, Wu T, Zeng L, Gong J. Nature of the Active Sites of VOx/Al2O3 Catalysts for Propane Dehydrogenation. ACS Catalysis 2016; 6 (8):5207-14.

[2] Sattler JJ, Ruiz-Martinez J, Santillan-Jimenez E, Weckhuysen BM. Catalytic dehydrogenation of light alkanes on metals and metal oxides. Chem Rev 2014; 114 (20):10613-53.

[3] Gong J, Zhao Z, Wu T, Xiong C, Sun G, Mu R, et al. Hydroxyl-mediated Non-oxidative Propane Dehydrogenation over VO_x/Al_2O_3 Catalysts with Promoted Stability. Angew Chem Int Ed Engl 2018;Advance online publication.

[4] Zhao H, Song H, Chou L, Zhao J, Yang J, Yan L. Insight into the structure and molybdenum species in mesoporous molybdena–alumina catalysts for isobutane dehydrogenation. Catalysis Science & Technology 2017; 7 (15):3258-67.

[5] An H, Zhang F, Guan Z, Liu X, Fan F, Li C. Investigating the Coke Formation Mechanism of H-ZSM-5 during Methanol Dehydration Using Operando UV–Raman Spectroscopy. ACS Catalysis 2018; 8 (10):9207-15.

[6] Moncada J, Adams W, Thakur R, Julin M, Carrero C. A, Developing a Raman-spectrokinetic approach to gain insights on the structure-reactivity relationship of supported metal oxide catalysts, ACS Catalysis; 2018; 8, 8976

573506 Investigation of Chabazite and Other Small-Pore Zeolites in the DME Carbonylation Reaction

Carbon-carbon bond formation via carbonylation reactions constitutes an important step in upgrading low-value feedstocks derived from syngas and biomass into useful chemical intermediates. The carbonylation of dimethyl ether (DME) is one such reaction, that occurs in the presence of acid catalysts to yield methyl acetate (MA), an industrially important solvent and platform chemical. MA production activity in mordenite (MOR) and ferrierite (FER) zeolites is attributed to the 8-membered ring (8MR) side pockets of these topologies, that facilitate the rate-limiting attack of CO onto the methoxy intermediate due to an optimized geometry.

In this work, we investigated the activity and location of the active site of a small-pore, 8MR-containing zeolite topology, chabazite (CHA). We found that the activity towards MA production orders MOR > CHA > FER for comparable Si/Al (~10), and that CHA-type zeolites (SSZ-13s) synthesized with $5 < Si/Al < 70$ exhibit an activity maximum centered near Si/Al = 10 that does not scale with total acid site density. To understand this, we conducted high-spin ^1H MAS NMR analyses on dehydrated SSZ-13s and divalent cation-exchange experiments to develop a structure-activity model for the active site in the CHA framework. The promising activity of CHA motivated the investigation of other small-pore, 8MR-containing topologies (e.g., LEV, AEI) as catalysts in the carbonylation of DME. Understanding the origin of activity in these types of frameworks may lead to applications in other carbonylation chemistries of interest.

575468 Competitive Transient Electrostatic Adsorption Mechanism for in Situ Regeneration of Poisoned Catalyst

Poisoning is identified as one major cause of catalyst deactivation, resulting in billions of dollars loss every year for catalyst replacement and process shutdown. Due to strong chemisorption of poisoning species (impurities in reactant feeds or generated during reaction), active sites would be occupied and deactivated. Thus regeneration of poisoned catalysts becomes essential to recover the catalytic activity to enable their reuses. Till to date, off-site catalyst regeneration remains a common procedure in practice. However, off-site regeneration process requires periodic interruption of reaction for taking out deactivated catalyst materials for regeneration and replacing them with fresh ones, which often adds complexity over reaction control and additional operation cost. To overcome this issue, in situ catalyst regeneration is demanded.

Herein we report one new Competitive Transient Electrostatic Adsorption (CTEA) mechanism for energy-efficient and noninvasive in situ regeneration of poisoned catalyst. We utilize the Townsend discharge phenomena for Ar^+ generation at a moderate DC voltage, and utilize CTEA of Ar^+ for competition against chemisorbed poisoning species to regenerate the active sites. We verified effectiveness of this new concept by studying HCOOH decomposition as one model reaction and examining the effects of CTEA of Ar^+ on the reaction properties. By applying a moderate DC voltage to create an electric field that generates Ar^+ ions following the Townsend discharge mechanism, the deactivated Pt catalyst exhibited an immediate recovery of the activity. The extent of the catalyst activity recovery was discovered to increase proportionally to the measured Ar^+ current, which was attributed to a larger number of Ar^+ ions that facilitate the desorption of chemisorbed CO poisoning species to regenerate the active sites via the CTEA mechanism. DFT simulations together with Townsend discharge theory suggested that the electrostatic adsorption energy of Ar^+ ions was dramatically bigger than the desorption energy of poisoning CO, which would force CO desorption to recover availability of the Pt active sites. The findings suggest CTEA mechanism offers a new, convenient and effective method to in situ regenerate poisoned catalyst materials, and shows a good potential to help reduce the operation cost associated with poisoned catalyst replacement in many reaction processes.

570559 Understanding the Performance and Limitations of Membrane-Electrode-Assembly Architectures for the Electrochemical Reduction of CO_2

The electrochemical reduction of CO_2 (CO_2R) to value-added products is an attractive technology for tackling the rising atmospheric CO_2 levels and storing intermittent renewable energy in chemical bonds. To achieve commercially-relevant CO_2R rates (> 100 mA cm^{-2}), gas-diffusion electrodes (GDEs) play an important role as they overcome mass-transport limitations that result from unfavorable CO_2/OH^- interactions and large diffusion lengths observed in aqueous, planar systems.[1] The thin diffusion layer in GDEs, typically in the order of 10 to 100 nm, also allows operation of CO_2R in alkaline conditions, which have been shown to suppress the hydrogen evolution side reaction (HER), and reduce the onset potentials of CO_2R products such as CO and C_2H_4.[2-5] With the increase in total current density, conventional cell designs for planar electrodes become severely limited by the ohmic drop across the cell, making membrane-electrode assemblies (MEAs) an attractive alternative. MEAs have smaller cell resistances as they do not have aqueous electrolyte compartments and minimal distances between the two electrodes.

In this talk, we present a multiphysics modeling framework for MEA systems performing CO_2R. We discuss the performance and limitations of two MEA designs: one with gaseous feeds at both the anode and cathode (full-MEA), the other with an aqueous anode feed (KHCO
and a gaseous cathode feed (exchange-MEA). Finally, we draw insights from sensitivity analysis of key parameters such as catalyst-layer loading and porosity, membrane thickness, kinetic rate constants etc. to guide the design of next-generation CO_2R devices and elucidate knowledge gaps to inform future research areas.

Acknowledgements
This material is based upon work performed by the Joint Center for Artificial Photosynthesis, a DOE Energy Innovation Hub, supported through the Office of Science of the U.S. Department of Energy under Award Number DE-SC0004993. We thank David Larson for providing experimental data for CO_2 reduction on membrane-electrode assembly devices.

1. T. Burdyny and W. A. Smith, *Energy Environ. Sci.*, 2019, DOI: 10.1039/c8ee03134g.
2. S. Verma, X. Lu, S. Ma, R. I. Masel and P. J. Kenis, *Phys. Chem. Chem. Phys.*, 2016, **18**, 7075-7084.
3. C. T. Dinh, T. Burdyny, M. G. Kibria, A. Seifitokaldani, C. M. Gabardo, F. P. G. de Arquer, A. Kiani, J. P. Edwards, P. De Luna, O. S. Bushuyev, C. Q. Zou, R. Quintero-Bermudez, Y. J. Pang, D. Sinton and E. H. Sargent, *Science*, 2018, **360**, 783-787.
4. C.-T. Dinh, F. P. García de Arquer, D. Sinton and E. H. Sargent, *ACS Energy Lett.*, 2018, **3**, 2835-2840.
5. D. Raciti, M. Mao, J. H. Park and C. Wang, *J. Electrochem. Soc.*, 2018, **165**, F799-F804.

571124 Aqueous-Phase Hydrogenation of Aromatic Rings and Carbonyl Compounds in the Presence of Applied Potential

Increasing availability of sustainable electricity and its intermittent character is stimulating the interest in storing electrical energy into chemical bonds. In this work, we aim to couple electrolytic H_2 production with chemical transformations of organic compounds aiming valuable chemicals and fuels. Thus, a series of carbon-supported metals have been tested for the electrocatalytic reduction of phenolic compounds, ethers, and aldehydes in aqueous conditions.

Electrocatalytic hydrogenation (ECH) of series of phenolic compounds, di-aryl ethers and aromatic aldehydes, has been performed in C-supported metals as working electrodes (cathodes). H_2 evolution (HER) is the prevalent side reaction on all metals, thus we define efficiency as the fraction of electrons used to reduce the organic compounds.

The rates of ECH of all tested compounds, as well as HER, increase with increasingly negative potentials. However, the efficiency is a strong function of the metal. Pt, for instance, strongly favors HER leading to low efficiencies, whereas Pd is highly efficient for reduction of carbonyl groups in aldehydes.

In the conversion of substituted phenols and diaryl ethers, the dominant reaction pathway is hydrogenation to cyclic alcohols and cycloalkyl ethers. Methoxy and benzyloxy groups undergo C-O bond cleavage via hydrolysis (minor route) and hydrogenolysis, which opens pathways to the formation of hydrocarbons. In stark contrast, aromatic aldehydes and ketones undergo only reduction of the carbonyl groups without exhibiting any hydrogenation of the aromatic ring. Interestingly, the presence of aromatic alcohol and acids during the conversion of aldehydes significantly increases the conversion rates of the carbonyl group, which is related to proton donation towards the carbonyl group during hydrogenation.

The results show that electrocatalysis drives the desired transformations at mild conditions and allow accurate control over the reaction pathways. Strategies to increase reaction rates at these conditions are increasing the concentration of hydronium ions and fine tuning the particle size of the metal. These effects are discussed in the light of competitive adsorption of hydrogen and the organic on the metal as key parameter determining the delivery of reducing equivalents, proton reduction and organic reduction.

577065 Development of Pt-Based Electrocatalysts for the Upgrading of Biomass-Derived Compounds Via the Kolbe Reaction

In recent years, there has been growing interest in the upgrading of biomass-derived molecules through an electrochemical method to produce biofuels and value-added chemicals. As compared to classical thermal-catalytic upgrading processes, the electrocatalytic process can be performed at low temperatures and pressures (293 K and 101 kPa) without the need to supply hydrogen. Whereas aldehydes, ketones, and phenolic compounds can be electrocatalytically reduced, carboxylic acids cannot be reduced.

Among all electrochemical processes, Kolbe electrolysis is the one of the oldest and best-known process for electrocatalytic conversion of carboxylic acid to alkanes and olefins via decarboxylation. However, the high current and electrode stability requirements combined with the complex reaction mechanism (including non-Kolbe) and low Faradaic efficiency has limited its application.

Herein, we reported the use of Pt-based electrocatalysts for Kolbe electrolysis. The effect Pt structure, supporting materials, electrolyte composition, and applied potential on the selectivity and Faradaic Efficiency of valeric acid oxidation were investigated. As shown in Figure 1, the valeric acid could be oxidized to octane (Kolbe), butene (Kolbe), and butanol (non-Kolbe) and their corresponding selectivities could be controlled by tuning the electrolyte composition and the applied potential.

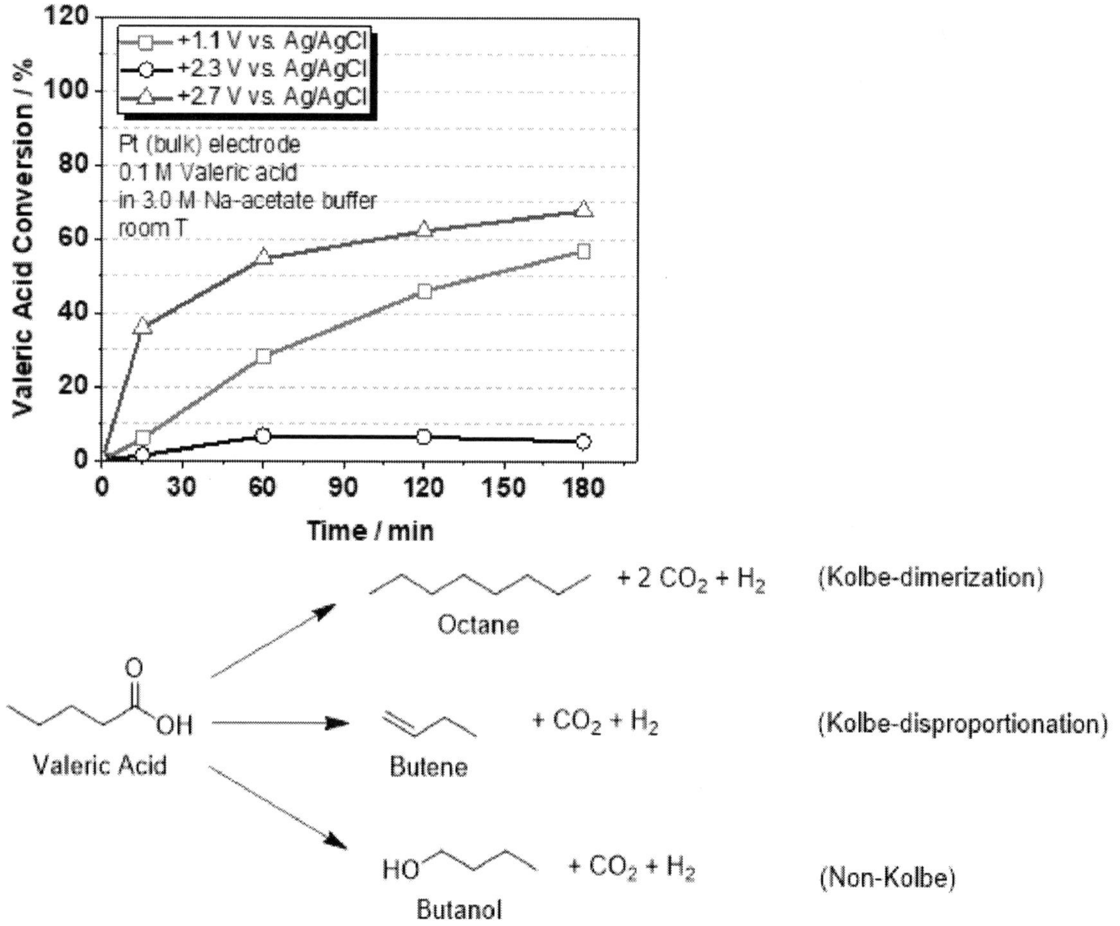

Figure 1. Electrochemical oxidation of valeric acid on Pt-based electrocatalysts under different applied potentials (left); and major reaction pathways (right).

576515 Chemical Transformations Via Electrochemical Reduction of Organics to Products

In recent years there has been growing interest in using electrochemical methods for hydrogenation of biomass-derived molecules for the formation of biofuels. Conventional hydrogenation processes require moderate temperatures between 433 and 678 K, pressures of 14,000 kPa, and external sources of dihydrogen for the stabilization and formation of biofuels. However, the same reaction can be performed using electrochemical reactors and much lower temperatures and pressures (293 K and 101 kPa) and with no supplied molecular hydrogen.

We have been exploring electrocatalytic upgrading of bio-oils. Our initial tests successfully decreased the carbonyl and aromatic species; however, the fundamental chemistry was not well understood. In this work, we use a combined experimental–theoretical approach to identify atomistic features and parameters that affect the electrochemical hydrogenation (ECH) rates of aldehydes, ketones, and carboxylic acids which are typically found in biocrude. In particular, we evaluated Pd, Rh, Ru, Cu, Co, Ni, and Zn metals supported on a carbon foam for the organic ECH and correlated the activity with computationally-derived parameters. We then used these parameters to explain the dependence of molecule functionality on the ECH rates. Our results show that whereas the reduction of aldehydes and ketones into alcohols can be achieved electrocatalytically at normal temperature and pressure, carboxylic acids cannot be reduced. However, carboxylic acids can be oxidized to alcohols, olefins, and ketones via Kolbe and non-Kolbe mechanism. Further, we observed there the turn over frequency (TOF) and faradaic efficiency (FE) depend on metal, fractional exposure, substrate concentration and half-cell, working vs reference (WvR), potentials. For example, as shown in Figure 1, the FE and TOF ECH are the highest when using Pd electrodes with low Pd weight loading (and small metal particle size). The FE and TOF ECH decreases as the Pd weight loading (and particle size) increases. Eley-Rideal-type kinetics were used to understand the structure sensitivity of the Pd electrocatalysts and revealed that it is related to changes in rate constants and substrate surface coverage. That is, small Pd nanoparticles have higher ECH rate constant and higher benzaldehyde surface coverage compared to large Pd particles. Structure sensitivity was also observed on Pt and Rh-based electrocatalysts, but with opposite trend. That is, large Pt and Rh nanoparticles were more active than small nanoparticles. Cu-based catalysts were structure insensitive.

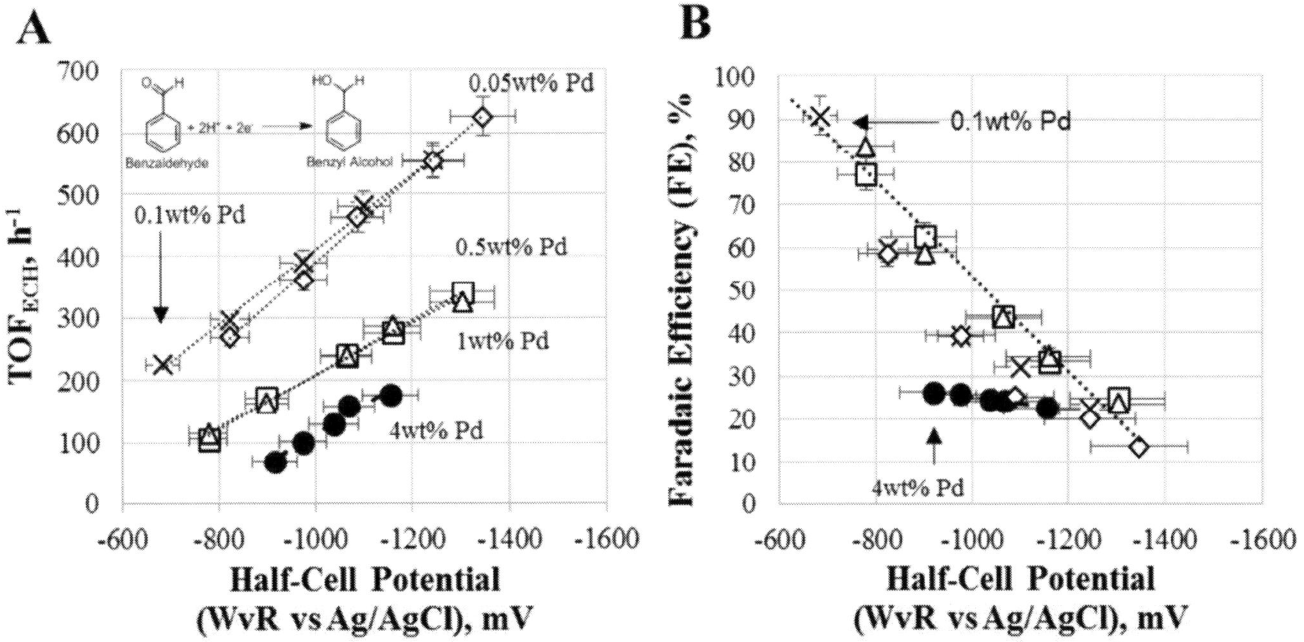

Figure 1. TOFECH (A) and Faradaic Efficiency (B) as a function of WvR potentials under normal conditions for the benzaldehyde ECH to benzyl alcohol. Catalysts with different Pd loadings were used where represents 0.05wt% Pd/CF, represents 0.10wt% Pd/CF, represents 0.50wt% Pd/CF, represents 1.0wt% Pd/CF, and represents 4.0wt% Pd/CF.

To better understand the reactions, we used computational theory. Classical molecular dynamics simulations were performed on the metal's most stable surfaces. All atoms of the electrode were fixed in position. The GROMACS package(83) was used for all classical molecular dynamics (MD) simulations. The force field's nonbonded interaction (van der Waals and electrostatics) parameters for classical MD simulations were fit to match the DFT (with Grimme's D3 correction(72)) binding energies to appropriately incorporate short-range bonded and long-range nonbonded interactions among the organic, solvent, and electrode molecules. Our computational model showed that the differences in catalytic activity were most likely due to differences in binding energy. The results followed the Sabatier principle which indicated that for the highest activity, the substrate must bond strong enough to be activated, but not so strongly that it cannot be released after reaction (Figure 2). Therefore computational modeling using our approach can be used to guide metal selection for the desired reactant reduction.

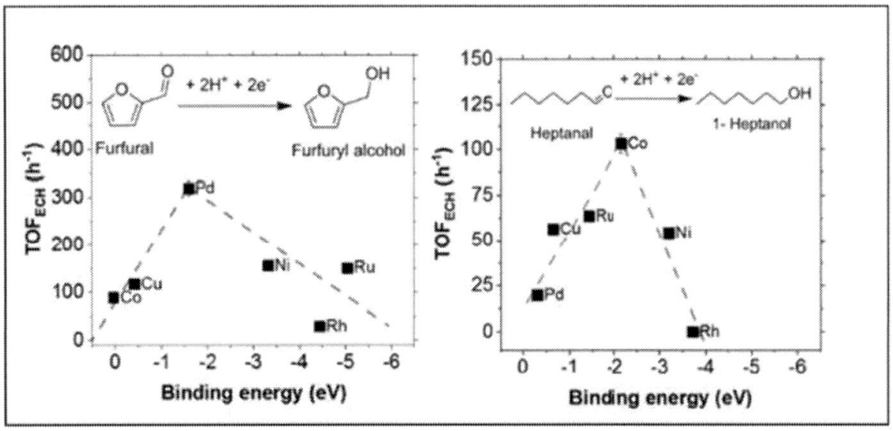

Figure 2. Turn over frequency vs binding energy of different metal catalysts with furfural and heptanal

577626 Assessing the Feasibility of Conceptual Electrocatalytic Processes through Inverse Techno-Economic Modeling

As the price of renewable electricity approaches that of fossil electricity, there are emerging opportunities to harness this low-cost, sustainable, electrical energy to decouple carbon emissions from economic development. Electrochemical processes are poised to play a pivotal role in the evolving energy ecosystem as the efficient interconversion of electrical and chemical energy can enable the deployment of sustainable technologies that support decarbonization of the electric grid, power the automotive fleet, and offer new opportunities for chemical manufacturing. With several notable exceptions, electrosynthesis is not yet widely employed in the industrial sector, but such processes have the potential to reduce plant cost through modularity and process intensification, improved safety through ambient operations, and increased process flexibility. Indeed, electrochemical processing may allow new molecular transformations, unlocking previously inaccessible or unimaginable synthetic routes. However, as compared to thermochemical processes, our understanding of the catalyst science and reaction engineering of electrochemical transformations is limited to a narrow set of reactions, largely influenced by historical interest in fuel cells and batteries. As such, while there is growing interest, there are few demonstrations of technological feasibility and these concepts remain nascent.

As economic considerations are a primary driver to technology adoption, we have developed a generalized techno-economic model that connect system performance and cost targets to the physical, electrochemical, and cost parameters of constituent components with a broader goal of quantitatively assessing a diverse set of conceptual electrochemical systems. By integrating this model into early-stage fundamental and applied research efforts, we articulate performance benchmarks for multiple components, highlight gaps in knowledge, and, ultimately inspire innovative application-informed approaches. In this presentation, I will describe model formulation and applications using three different case studies. First, to establish the accuracy of such a model, I will evaluate water electrolysis as a means of hydrogen production comparing our results to established U.S. Department of Energy models. Second, I will assess the feasibility of electrochemically upgrading biomass-derived small molecules for use in fuel or chemical manufacturing. Third and finally, I will illustrate that while, depending the catalyst choice and reaction conditions, carbon dioxide electroreduction can yield many scientifically-interesting products, only a subset are viable technology options. Such inverse techno-economic models can guide research investigations to address the most important limitations in proposed technology platforms and to provide credible pathways to cost-effective deployment.

574774 Electrochemical Oxidative Dehydrogenation (e-ODH) As a Process Intensification Platform in Shale Gas Upgrading

The current availability of Shale Gas has resulted in an increase in ethane availability due to its higher content of natural gas liquids (NGLs) in comparison to conventional natural gas. The majority of the new NGL capacity comes from natural gas producing plays (Utica and Marcellus shale) and associated gas from tight oil production (Eagle Ford and Bakken shale). However, ethane is costly and energy intensive to recover from natural gas as cryogenic (turbo-expansion) processing is required. As such, this ethane oversupply leads to a rejection of ethane to the gas pipeline system to eliminate the recovery cost. Still, this technique can not completely manage the ethane overabundance as the gas condensation may occur through transportation under operating pressure. Although the NGLs export is another management technique, most of NGLs plants are far from the coasts.

Within this context, Ohio University (OHIO), with funding from the U.S. Department of Energy's National Energy Technology Laboratory (NETL) Project DE-FE0031709, is developing a modular electrogenerative oxidative dehydrogenation (e-ODH) process, which directly converts NGLs at the well head into fuels, electrical power, and pipeline-quality natural gas. In this process, ethane (and possibly other NGLs) contained in the well head gas are selectively converted into alkenes and byproduct electrical power using a solid oxide fuel cell (SOFC) module. This module could possibly be coupled with an oligomerizaton reactor for additional alkene upgrading. Advantages offered by this e-ODH process include the following: 1) Modular operation with lower capital and operating costs; 2) Selective conversion of NGLs contained in well head gas; 3) Production of gasoline range hydrocarbons, pipeline-quality natural gas, and electrical power as products; and 4) Utilization of existing SOFC and oligomerization reactor technology minimizing commercial adoption and market entry risk.

OHIO has been using both process models and experimental analyses to develop the e-ODH process via Aspen Plus simulations and high temperature catalyst development, respectively. This presentation will discuss results from both initiatives during the meeting.

562023 Improvement of Transportation Fuels Sustainability Via Co-Hydroprocessing of Lipid-Feedstocks and Petroleum Fractions

Improvement of transportation fuels sustainability via co-hydroprocessing of lipid-feedstocks and petroleum fractions

Stella Bezergianni, Centre for Research & Technology Hellas, Thermi-Thessaloniki, Greece*

Athanasios Dimitriadis, Centre for Research & Technology Hellas, Thermi-Thessaloniki, Greece

Loukia Chrysikou, Centre for Research & Technology Hellas, Thermi-Thessaloniki, Greece

Vasilios Dimitropoulos, Hellenic Petroleum, Thessaloniki, Greece

Maria Maggiliotou, Hellenic Petroleum, Thessaloniki, Greece

Spyridon Kiartzis, Hellenic Petroleum, Thessaloniki, Greece

Biography

Dr. Bezergianni is a principal researcher at the Centre for Research & Technology Hellas, active in R&D of new environmentally friendly fuels and biofuels. She is involved in new biofuel production technologies regarding the valorization of bio-based feedstocks for the production of drop-in alternative fuels and as well as of hybrid fuels via the integration of bio-based feedstocks in refineries. Dr. Bezergianni has an extended academic (numerous scientific publications and monographs) and research record (scientific advisor of various European and national research projects and direct collaboration with national and international partners), as well as prior industrial experience in and out of Europe (ExxonMobil, FMC Corporation, Air Liquide). She is an active member National Technical Chamber of Greece and a committee member of the American-Hellenic Chamber of Commerce.

Abstract

Even though biofuels are promoted as one of the most significant substitutes of fossil fuels with a considerable share growth in recent years, they are still associated with high investment costs, uncertain environmental benefits and questionable compatibility with conventional combustion systems [1]. One alternative approach for promoting a drastic increase of biomass utilization for the production of greener fuels for the transportation sector is co-processing of bio-based feedstocks in conventional refining conversion units [2]. An R&D project, driven by the Centre for Research and Technology Hellas and Hellenic Petroleum on integrating waste cooking oil in a conventional heavy gasoil hydrotreatment unit, highlighted the technical feasibility as well as environmental and economic potential. The positive prospects are currently evaluated via a pilot scale test for a Hellenic Petroleum refinery. This presentation will focus on key R&D perspectives, challenges and potential for integrating residual lipids in conventional catalytic hydroprocessing units. The presentation will provide practical and useful information for the refiners, in order to enable a fast integration of biomass within existing conventional refineries for the production of high bio-content fuels.

Introduction

The proof-of-concept was initially addressed via the SustainDiesel research project funded by the Greek Secretariat for Research & Technology where both the technical and environmental potential benefits from integrating waste cooking oil were estimated (see Figure 1).

Properties	Units	Hybrid diesel	Market diesel
Density	kg/m³	825.8	829.2
Flash point	°C	61	67
Sulpur	wppm	8.2	5.8
Viscosity (40°C)	cSt	2.996	3.066
Cetane index		59.1	58.3
Cetane number		56.7	55.7
Water	wppm	100	60
CFPP	°C	-2	0
Polyaromatics	%w/w	1.1	1.4
Lubricity	μm	181	174
Recovery 95% v/v	°C	353.0	353.2
Recovery at 250°C	%v/v	26.8	25.8

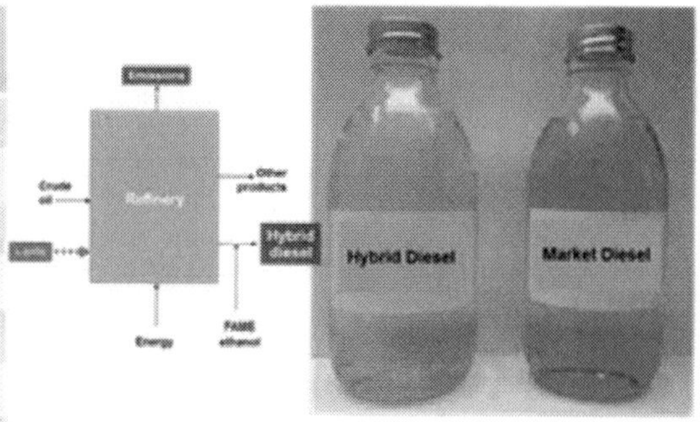

	Hybrid diesel ($kg\ CO_2$-eq/l)	Market diesel ($kg\ CO_2$-eq/l)
WTT analysis	0.40	0.45
Processing in refinery	0.37	0.36
FAME	0.03	0.09

Co-feeding 5% WCO ⇒ 11% CO_2 emissions reduction

www.sustaindiesel.gr

Figure 1. Overview of SustainDiesel project results [3-5]

The positive research results have driven Hellenic Petroleum to explore further this technology, via a self-funded project on exploring the potential co-hydroprocessing of waste lipids in underlying hydrotreating units.

Methodology

The project involved a) an evaluation of valiable lipid feedstocks for co-feeding with heavy gas oil, b) an assessment of the optimal lipid management strategy within refinery, and c) an extended pilot testing of co-hydrotreating heavy gas oil with lipids for hybrid fuels production validation.

Evaluation of available lipid feedstocks for co-feeding with heavy gas oil

Three tests were performed for the evaluation of the effect of different lipid type (sunflower oil, raw/crude waste cooking oil and refined waste cooking oil). For each test, three phases were considered, a nominal phase 1 where the heavy oil defined the basic hydrogenation activity, a main phase 2 where the lipid feedstock was introduced aiming to obtain a product with <10 wppm sulphur content and a third phase where only heavy oil was used that was intended to identify any permanent loss of activity for the employed catalytic system.

In Figure 2 the results of the tree tests are outlined indicating that the least impact was for sunflower oil and refined waste cooking oil. The slightly largest impact of waste cooking oil can be expected as this lipid source contains unsaturated oxygenated hydrocarbons as well as free fatty acids. The permanent loss of the catalyst effectiveness is also more clearly indicated for raw waste cooking oil, which can be primarily attributed to the free fatty acids, as well as on the negative effect of the CO_2 and CO (by-products of co-hydroprocessing resulting from hydro-deoxygenation reactions) on the catalyst active sites [6-7].

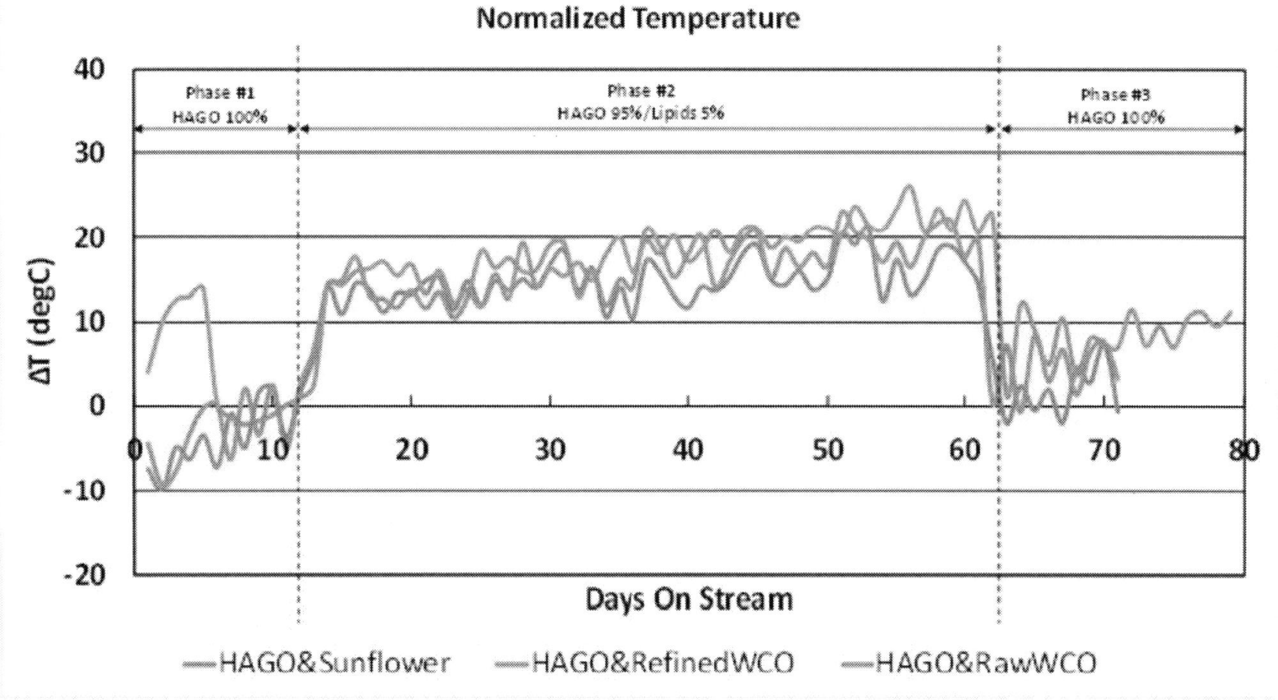

Figure 2. Normalized reaction temperature indicating hydrotreating effectiveness for suflower oil, raw waste cooking oil and c) refined waste cooking oil co-hydroprocessing with heavy gas-oil

Assessment of the optimal lipid management strategy within refinery

The evaluation of the potential logistics for handling lipid feedstocks within a refinery complex was performed via extended storage studies that evaluated the sunflower oil and refined waste cooking oil 6-12 month storage. As the lipid feedstocks have a tendency to oxidize that leads to further water formation, the water content and acidity were monitored throughout the extended storage studies. The studies have shown that for a 6-month storage period the water content was not significantly changed however the acidity of sunflower oil increased as indicated in Figure 3. As the water content and acidity of the lipid source has been identified as potential causes of the catalyst hydrogenation activity limitation affecting the catalyst active sites, the consideration of the 6-mont storage as a maximum time-span for storing the lipid sources is of particular essence to the industrial application of this technology.

Extended pilot testing of co-hydrotreating heavy gas oil with lipids for hybrid fuels production

Upon the evaluation of the different lipid-sources with respect to the impact on heavy gas oil hydrotreatment as well as on the extended storage, a prolonged pilot testing was performed, validating the effectiveness of hybrid fuels production via lipid co-hydroprocessing with heavy gas oil. The testing was performed in the large scale pilot plant of CERTH for 6 months, proving the effectiveness of sustainability of the proposed technology.

The testing was conducted with refined waste cooking oil (5% v/v) with standard heavy gas oil, over a commercial $CoMo/Al_2O_3$ catalytic system and respecting the operating window of heavy gas oil hydrotreatment of the refinery. The testing provided evidence that the technology can be scaled up to industrial plants, expecting high quality fuels and minimal additional energy expectations.

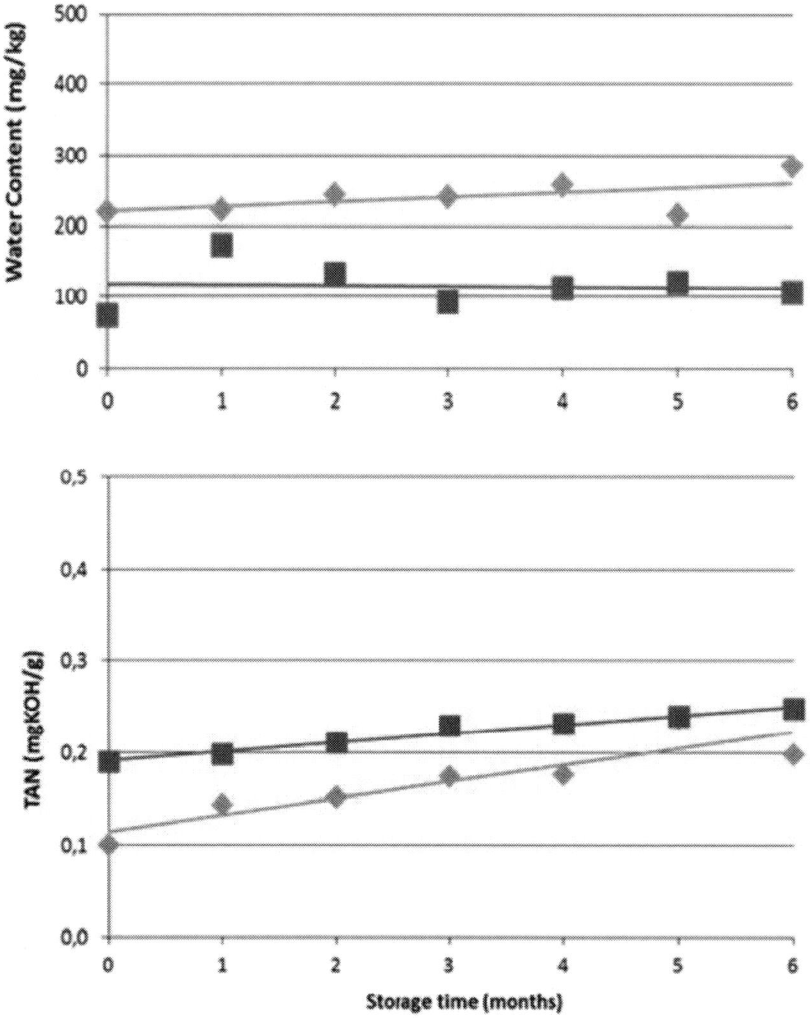

Figure 3. Water content and Total Acid Number (TAN) of sunflower oil (orange) and refined waste cooking oil (red) for an extended 6-month storage period

Conclusions

The integration of lipid sources (primarily refined waste cooking oil and raw vegetable oils) within an underlying hydroprocessing plant of a refinery is feasible and renders high quality products. The additional H_2 required for co-hydroprocessing lipid feedstocks up to 5% are not significant and the resulting hybrid fuels are of high quality and abide by the fuel standards.

References

[1] A. Dimitriadis, I. Natsios, A. Dimaratos, D. Katsaounis, Z. Samarasm S. Bezergianni, K. Lehto. Frontiers in Mechanical Engineering (2018) https://doi.org/10.3389/fmech.2018.00007

[2] Bezergianni, S., Dimitriadis, A., Kikhtyanin, O., Kubicka, D. Refinery co-processing of renewable feeds, Progress in Energy and Combustion Science 68, pp. 29-64, 2018

[3] L.P. Chrysikou, V. Dagonikou, A. Dimitriadis, S. Bezergianni. Journal of Cleaner Production 10 (2019) 566-575

[4] Bezergianni, S., Dimitriadis, A., Karonis, D. Diesel decarbonization via effective catalytic Co-hydroprocessing of residual lipids with gas-oil, Fuel, 136, pp. 366-373, 2014

[5] Bezergianni, S., Dimitriadis A, Meletidis, G., Effectiveness of CoMo and NiMo Catalysts on Co-Hydroprocessing of Heavy Atmospheric Gas Oil - Waste Cooking Oil Mixtures, Fuel, 125, pp. 129-136, 2014

[6] S. Bezergianni, V. Dagonikou. Journal of Chemical Engineering 93 (2015) 1017-23

[7] S. Bezergianni, V. Dagonikou, S. Sklari. Fuel Processing Technology 144 (2016) 20-6.

-

563626 Pyrolysis of Solid Food Court Waste Mixtures Coupled with *Ex-Situ* Catalytic Upgrading

Pyrolysis of Solid Food Court Waste Mixtures Coupled with Ex-situ Catalytic Upgrading

Gabriela Ionescu[1], Razvan State[1], Dorin Boldor[1,2], Cosmin Marculescu[1]

[1] University Politehnica of Bucharest, 060042, Romania

gabriela_ionescu@ymail.com, razvan.state@yahoo.com, cosminmarcul@yahoo.co.uk,

[2] Louisiana State University Agricultural Center, Baton Rouge, 70803, USA

dboldor@agcenter.lsu.edu

Abstract: The use of renewable resources as bioenergy feedstock has become an intensely studied area towards sustainability issues and climate change mitigation. Their usage could contribute to the reduction of greenhouse gas emission by 2050 with 50-80% [1]. The European Union (EU) already started its transition towards a bio-based economy by conceiving a circular economy strategy that implies the following main targets by 2030: maximum 10% of all waste produced should be landfilled, while 70% of municipal waste and 80% of packaging waste recovered [2]. The structure and properties of renewable feedstocks such as woody, non-woody biomass and light packaging waste allows their conversion into valuable energetic products in gas, liquid (bio-oil) and solid (bio-char) form. In the last decade, the thermal cracking process performed in oxygen free atmosphere (pyrolysis) has been investigated for production of various fuels (from both fossil and non-fossil resources). Parameters studied include type of feedstock (neat or in mixture – co-pyrolysis), different process designs and parameters, with or without catalysts, as stand-alone process or as an intermediate one [3,4]. Depending on the type of renewable feedstock, the distribution of the by-products can vary between: 13-35% for pyrolysis gas, 30-80% for bio-oil, and 2-35% for bio-char [5]. Due to their significant calorific values they can be used as low-grade fuels or upgraded using catalysts, for obtaining drop-in fuels.

This study aims to investigate the effects of ex-situ catalytic upgrading of volatile matter resulted from solid waste mixtures pyrolysis, generated in a tubular batch reactor at 500°C , and two different catalysts heated at 350°C and 400°C. The experimental feedstock consists of a mixture of organic and light packaging residual waste collected from food court areas waste. To estimate its average composition, the solid waste mixtures were collected in diverse seasons, from different shopping malls, located in Bucharest, Romania. According to the American Society for Testing and Materials (ASTM) standard method E 871-82 the collected solid waste mixture samples were dried at 105°C for 24 h. The equivalent composition of the feedstock in dried basis resulted as follows: paper (29.2%), cardboard (26.2%), organic waste (25.2%), plastic (18.92%) and wood (0.4%). Subsequently, the dried feedstock was reduced in size, with an average particle diameter from the order of microns up to 2 mm. The experiments were carried out in a custom-made tubular electrically heated batch reactor (model RO 60/750/13 modified, NABERTHERM), in which a ceramic crucible was placed along with the feedstock. For each experiment 15 g of samples were distributed in the ceramic crucible, the furnace temperature being set constant at 500°C. The choice of the process temperature is based on our previous studies, which concluded that this optimal temperature for light packaging waste obtains higher gas and oil yields [6]. Before and during the process, 1 l/min of nitrogen (N_2) was introduced in front of the reactor. The catalytic reactor consists of a cylindrical quartz tube covered with heating tape. For each experiment 5 g of Fe^{2+}HZMS-5 and Fe^{3+}HZMS-5 zeolite type catalysts were place in the heated ceramic tube at 350°C and 400°C, respectively. The laboratory catalysts were prepared by wet impregnation of the iron precursor onto the zeolite support. During the process, the ceramic crucible was coupled with the catalytic system that allowed the pyrolysis gas to pass through the heated zeolite catalysts. Subsequently, the pyrolyzed gaseous species were carried out through a cooling system that permitted the collection of the condensable fraction. The non-condensable gaseous fraction composition was determined using a micro-gas chromatograph (Micro-GC Fusion, Inficon). The gases were extracted at the ending line of the cooling system using suitable syringes, throughout the entire pyrolysis process at different establish times. Then the syringes were injected in the Micro-GC gas analyzer, that allowed the detection of H_2, O_2, N_2, CH_4, CO, CO_2, C_2H_6, C_3H_6 C_3H_8, C_4H_6, C_4H_{10}, C_5H_{12}.The bio-oil and bio-char were deposited in recipients for subsequent experiments. All the tests have been repeated.

From the experimental results, the average yields distribution is presented as follows bio-char/bio-oil/permanent gases: for Fe^{2+}-HZSM-5- 28%/38%/33% at 350°C and 24%/38%/38% at 400°C, while for Fe^{3+}-HZSM-5- 28%/36%/36% at 350°C and 25%/29%/46% at 400°C.

For all the studied temperatures, the amount of O_2 present in the material and in the unoccupied volume of the pyrolysis reactor (some of it due to probably minor leaks), catalyst reactor and condensation system, decreases with the process evolution and stabilization, being totally consumed after approximately 4 minutes. The O_2 release, favors the CO_2 and CO formation, reaching to their maximum peak after the full release of the oxygen, and decreases with the residence time, tending to zero. In the devolatilization stage, nitrogen (from the purging gas) reaches to its minimum values due to the formation of carbon-based gaseous species, then slowly increases with the advancement of the process, as fewer volatiles are produced.

The pyrolysis gaseous species composition shows that the CO_2, CO and CH_4 are the primary products resulted from the process (Figure 1), followed by H_2, acetylene-ethylene (C_4H_6), ethane (C_2H_6), propane (C_3H_8), butane (C_4H_{10}), pentane (C_5H_{12}) (Figure 2). Overall, Fe^{2+}-HZSM-5 zeolite catalyst enhances the formation of useful gases at both temperatures studied (350°C and 400°C), reaching to a maximum of 16% for the major components and almost 2% for minor ones, at 400°C.

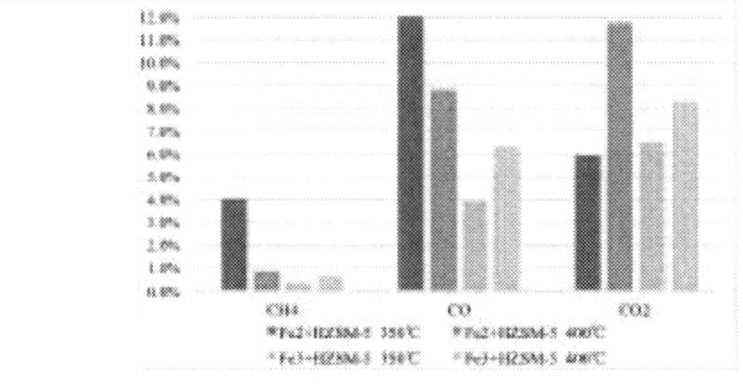

Figure 1. Average gas yields for major components evolved

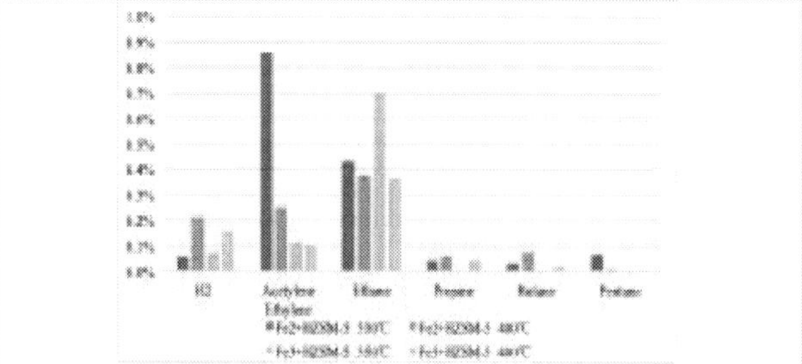

Figure 2. Average gas yields for minor components evolved

In conclusion, the experimental results revealed that Fe^{2+}-HZSM-5 catalyst registered better performance yields regarding the bio-oil production at both temperatures (350°C and 400°C) in comparison with Fe^{3+}-HZSM-5.

Keywords: *catalytic pyrolysis, biomass, plastic, waste, HZSM-5.*

Acknowledgement

Funding provided through Romania's "Competitiveness Operational Programme 2014-2020", Priority Axis 1: Research, Technological Development and Innovation (RD&I) to Support Economic Competitiveness and Business Development, Action 1.1.4. Attracting high-level personnel from abroad in order to enhance the RD capacity. ID / Cod My SMIS: P_37_768/ 103651; No. Contract: 39/02.09.2016.

References

1. J. Bogner, R. Pipatti, S. Hashimoto, C. Diaz, K. Mareckova, L. Diaz, P. Kjeldsen, S. Monni, A. Faaij, Q. Gao, Waste Manag. Res., 26 , pp. 11-32, (2008)
2. M. Ragazzi, S. Fedrizzi, E. C. Rada, G Ionescu,R. Ciudin, L. I. Cioca, Energy Procedia, 119, 192-200. (2017). 10.1016/j.egypro.2017.07.068
3. V. Dhyani, T. Bhaskar, Renewable Energy, 129, 695-716, (2018). 10.1016/j.renene.2017.04.035
4. R. State, A. Volceanov, M. Pranjali, D. Boldor, Bioresource Technology In Press, Accepted, (2019). doi.org/10.1016/j.biortech.2019.01.036
5. G. Ionescu, C. Bulmău, EENVIRO – 2018, 9-13th of October 2018, Cluj Napoca, Romania.
6. G.Ionescu, PhD thesis, University of Trento, University Politehnica of Bucharest, 2012.

577032 In-Situ Evaluation of the Concentration of 1-Hexene in a Laboratory-Scale Fluid Catalytic Cracking (FCC)

In-situ evaluation of the concentration of 1-hexene in a laboratory-scale Fluid Catalytic Cracking (FCC)

Juan Lacayo, Sebastián López, David Soto, Alejandro Molina.

jglacayol@unal.edu.co, selopezgo@unal.edu.co, dasotog@unal.edu.co, amolinao@unal.edu.co.

Bioprocesos y Flujos Reactivos, Facultad de Minas, Universidad Nacional de Colombia - Sede Medellín, Carrera 80 # 65-223, Medellín, Colombia

Presenting author: *Alejandro Molina.*

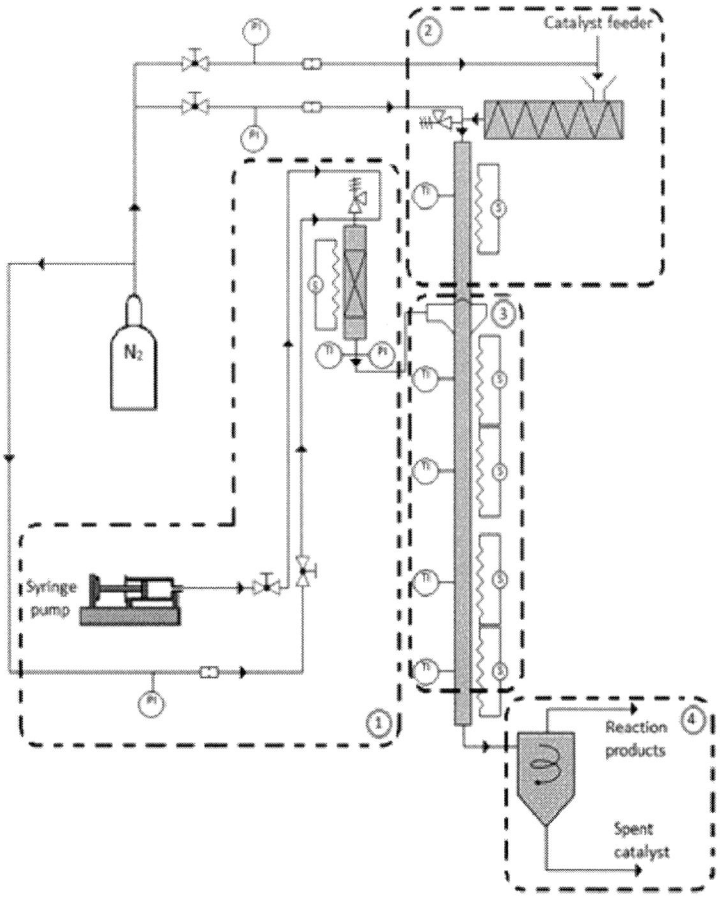

Figure 1. Experimental setup. 1) Reactant evaporation, 2) Catalyst heating, 3) Reaction, 4) Catalyst separation.

Abstract: The concentration of 1-hexene at conditions similar to those observed in FCC industrial units was determined in situ with an optical technique. Experiments were carried out in a laboratory-scale (internal diameter = 1.3 cm and 110 cm of length), semi-continuous downer reactor that allows optical access through fused quartz walls. Equilibrated FCC catalyst with a mean particle size of 84 μm was injected to the reactor at a rate of 30 g/min to a flow (300 - 3000 sccm) of 1-hexene that entered the system as a vapor at 373 K. The estimated catalyst to oil ratio was between 20 - 70. The light source was a helium-neon laser (2.0 mW minimum) operating at 2949.9 cm^{-1} (3.39 μm). The laser beam traversed the reactor and was modulated with an optical chopper operating at 500 Hz. A PbSe photodetector was coupled to a data acquisition system and the amount of light absorbed was quantified as fractional transmission (I/Io). Figure 1 presents a schematic of the reaction setup. Estimated residence times for particles and gas were between 2 - 5 and 8 - 25 seconds, respectively. The experiments were carried out in the range at temperatures varying between 373 K and 673 K and at 0.85 atm.

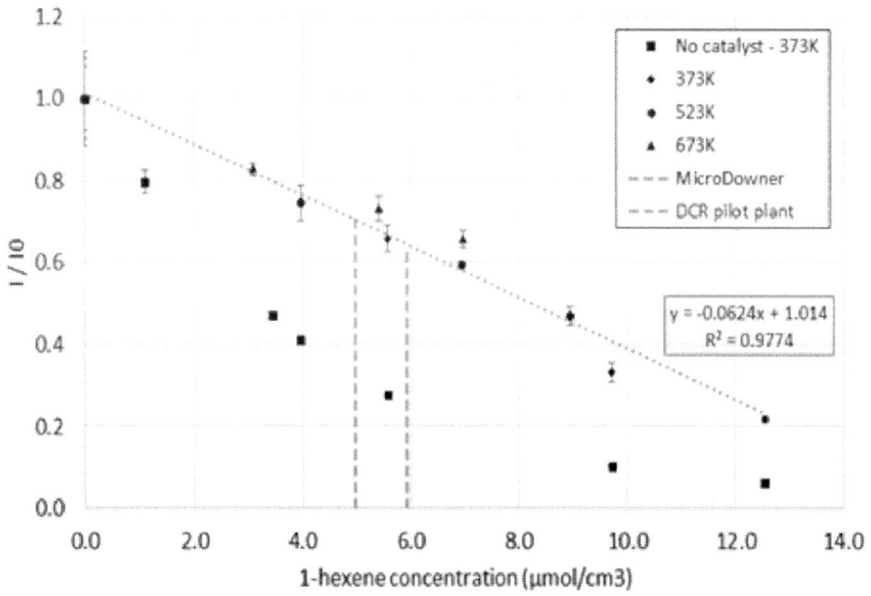

Figure 2. Fractional transmission (I/Io) at different 1-hexene concentrations and concentration of other experimental setups [1,2].

Figure 2 shows the variation of I/Io with the concentration of 1-hexene. The comparison is made for experiments in the absence and presence of the catalyst and at 373 K, 523 K, and 673 K. In the catalyst-free experiments, the fact that the variation of I/Io is not linear suggests that at high 1-hexene concentration the path becomes optically thick and linearity is lost. In the presence of particles, I/Io increases as the value of Io decreases due to light absorption by the particles. Nevertheless, a good linear correlation (r^2=0.9774) was found at all temperatures. Figure 2 also marks concentrations of experimental setups [1,2] used for the characterization of FCC systems. The fact that I/Io linearly correlates with the 1-hexene concentration at conditions similar to those used to characterize FCC and in the presence of catalyst gives confidence that this system can be used to measure the advance of the FCC reaction in industrial FCC units for process control applications and in the development of new petroleum refining technologies.

Keywords: Fluid catalytic cracking, downer reactor, in situ measurement.

References

reactor for the study of catalytic cracking. *Applied Catalysis A: General*, *232*(1-2), 247-263.

[2] Bryden, K., Weatherbee, G., & Habib, E. T. (2013). Flexible pilot plant technology for evaluation of unconventional feedstocks and processes. *Catalagram*, *113*, 3-21.

577182 Development of Rigorous Refining Simulation Models to Determine Greenhouse Gas Emission Impact of Canadian Crude Oil Bends in a Refinery

The change in crude oil blends significantly influences their properties and interactions in a refinery. Thus, crude oil blending is an important technical aspect that should be addressed for renewable transportation fuel production and reduced greenhouse gas in refineries. Previous research estimated refinery energy demand and/or greenhouse gas emissions using aggregated data and/or correlations, which are specific to a particular refinery or a crude blend. Results based on aggregated data and/correlations have limited accuracy. This study examines the properties and the interactions of five Canadian crude oil blends in a refinery using rigorous process simulation models. Data-intensive bottom-up models were developed to evaluate the energy and greenhouse gas emission profiles. Aspen Hysys simulation software was used to develop refinery process units and characterize crude oil feeds and product properties. The results of this study provides refining operators insights on how to obtain petroleum feedstocks that satisfy the desired product quality with reduced energy consumption and greenhouse gas emissions.

558627 Suitable Characteristics for Surfactants Substituent of Antifoam Silicon Oil

Petroleum is still the main source of energy around the world, according to the International Energy Agency (iea.org). For its great significance, much effort is applied on more effective ways to extract the oil and isolate it, decreasing the number of steps and even reducing the time required for it. Among the hindrances, the formation of foam is a concern because of the reduced oil production capacity observed, result of the large overheads in the gravitational separators where water, oil and gas are separated.

Among the methods for foam control, the more efficient is the application of antifoamers or defoamers, chemicals with properties to avoid foam to form or to break it in a short time. The standard product used in this case is silicone oil, specifically polydimethylsiloxane (PDMS). Although it is very efficient as an antifoam agent, the silicon present in PDMS structure is converted to silicon oxide in the petroleum refining step, causing the poisoning of the hydro-treatment catalyst and the yield reduction. Much research has been done to reduce the amount of silicon in the antifoamers, but none with satisfactory results[1]. Therefore, another approach has been made to understand what properties of PDMS make it a so effective antifoamer, and with those, been able to design a silicon-free polimer or a low-quantity silicon polimer with such characteristics.

In this manner, 5 Polydimethylsiloxane (PDMS) commercial samples with distinct viscosities had their viscosimetric molar mass determined. For each sample, 6 solutions of different concentrations were prepared in toluene and their viscosity was analyzed in a rotational rheometer Haake MARS 60, with shear rate defined at 300 s^{-1} at 25°C.The parameters K and a applied in the equation $[\eta] = K\ M^{a}$ were acquired from the literature[2]. Formulations of these samples were prepared and the interfacial tension with petroleum 19°API and the antifoam efficiency[3] were tested. Physical-chemical properties, such as dielectric constant, density at 20°C and 4°C, viscosity at 25°C and permeability of gas were also investigated. In order to determine the formulation with the most effective performance, different solvents were used in this study.

So far, it can be inferred that for samples with lower viscosity, or lower molar mass, the antifoam effect is almost null, despite been a silicon oil sample. With increasing the viscosity, the antifoam effect increases to a maximum and starts to decrease at even higher viscosities. As for this behavior, it is known that the higher the molecular weight, the greater the likelihood of entanglements, which may alter the dispersion of the surfactant, decreasing its antifoam potential.

Moreover, viscosity should be adjusted so the agent may be properly solubilized. For that, the solvent has a major role, as it may synergize with the agent so to improve its access to the oil, allowing a faster time of action avoiding the formation of the foam.

The interfacial tension allowed to observe how the samples modify the elasticity of the foam film by comparison of the entrance, spreading and bridge coefficients. The calculations shown a slight difference in spreading, indicating that the smaller molar mass samples access the oil medium easier. No changes were observed in the other 2 coefficients.

Surface rheology is being investigated, together with solubility parameters to understand how cohesion and adhesion of the antifoamers with the oil will affect the tensoactive behavior.

References:

1. Cevada, E. Roos K. Alvarez F. Carlotti S. Vázquez F. *Fuel*. **2018**, 221, 447-454.
2. Barry, A. J. J. *Phys*. **1946**, 70, 1,020.
3. Rezende, D. A. Bittencourt, R. R., Mansur, C. R. E. *Petrol. Sci. Eng*. **2011**, 76, 172-177.

556160 Impact of Deep Hydrogenation on Jet Fuel Oxidation and Deposition

Thermal oxidative stability is the main character for jet fuel serving as coolant of supersonic or hypersonic aircrafts. This study investigated the effect of hydrogenation degree on jet fuel (RP-3) oxidation and deposition. The deep hydrogenation of RP-3 were achieved by commercial noble catalyst in an autoclave with 4 MPa hydrogen at 190 °C, and the reaction time was controlled to acquire fuels with different hydrogenation degree. The aromatic content of the most hydrogenated fuel is reduced from 12.6% to 0.5%, and the total sulfur content is lowed to below 3 ppm. Accordingly, the density of hydrogenated fuels slightly decreases but the net heat of combustion slightly increases due to the saturation of aromatics. The oxidation reactivity and deposition tendency of hydrogenated fuels were further conducted by pressure differential scanning calorimetry (PDSC) and jet fuel thermal oxidation tester (JFTOT), respectively. The decrease of onset oxidation temperature and induction period of hydrogenated fuels highlight a significant effect of aromatic content on jet fuel oxidation. However, hydrogenated fuels exhibit lower deposition propensity upon thermal oxidative stress. These results indicate that deep hydrogenation can effectively improve jet fuel thermal oxidative stability (fewer deposits) but reduce the anti-oxidation effect of jet fuel. Oxidative products of aromatics and inherent heteroatomic compounds detected by comprehensive two-dimensional gas chromatography are responsible for the confliction impact of deep hydrogenation on oxidation and deposition.

569250 Biodiesel Production Using Biopolymer Encapsulated Lipase in Bioreactors

The aim of this study was to encapsulate lipase in a biopolymer and to produce biodiesel using the encapsulated lipase in an immobilized bioreactor. Encapsulated lipase was prepared using co-extrusion technique and κ-carrageenan as the matrix. This study has been undertaken due to environmentally benign, nontoxic, low cost, renewable characteristics of κ-carrageenan and uniqueness of co-extrusion technique. Lipase PS from Burkholderia cepacia was found to be efficient for bio diesel production among various lipases that were screened. Encapsulated lipase physico-chemical studies showed good catalytic, stability and reuse characteristics.Process parameters were optimized for biodiesel production. The optimum conditions for processing for biodiesel production in a stirred tank immobilized bioreactor and packed bed bioreactor were 30°C, 72 h reaction time, 23.7 x g relative centrifugal force and 1.5 ml/min and 264 h reaction time respectively. At the optimal conditions, biodiesel conversion up to 100% and 82% were achieved in stirred tank and packed bed reactor respectively. The life cycle impact of biodiesel from immobilized lipase as catalyst was less compared to the conventional catalytic processes. From the results, it is concluded that biodiesel production using encapsulated lipase in an immobilized bioreactor may open new outlooks for the scale-up studies of this technology in near future.

571292 KBR's Propane Dehydrogenation Technology (K-PRO) Meets Market-Driven on-Purpose Propylene Demand

Propylene is mostly produced worldwide as a by-product from either steam crackers or FCC units. However, with the large number of new ethane-based steam crackers coming on line in the next few years, the ongoing transition of many existing steam crackers from naphtha feedstock to ethane and propane feedstock, and with very few new FCC units being built, it is anticipated that these traditional sources of propylene supply will fall short of demand by 45 MMTPA in 2027. 80% of this propylene shortfall is expected to be made up by on-purpose propane dehydrogenation. KBR's Propane Dehydrogenation Technology (K-PRO^{TM}), based on the commercially proven KBR Catalytic Olefins Technology (K-COT^{TM}) in combination with a novel high selectivity dehydrogenation catalyst, is well positioned to fill this gap. The technology can be implemented as a stand-alone propylene production unit independent of a steam cracker or an FCC unit.

During this presentation, we will present the details of K-PRO^{TM} Technology and discuss its benefits including high propylene selectivity and conversion, low-cost and environmentally friendly catalyst, minimal catalyst losses and catalyst make-up requirements, how its commercially proven reactor design leads to simple and reliable operation with high on-stream factors, and lower CAPEX and OPEX as compared to other technologies.

572699 Process Conceptualization in a Carbon-Constrained World: Leveraging Optimization and Systems Engineering to Balance Multiple Objectives

The search space for chemical process design can be remarkably complex and typically involves an interplay of both integer (e.g. what unit operations to consider?) and continuous decision variables (e.g. what operating conditions to run at?). Even for a seemingly simple unit operation like distillation, the number of potential configurations scales exponentially with the number of product streams (>80 million configurations for 6 products). As a result, it is no surprise that even seasoned practitioners rely heavily on heuristics and rules of thumb to "lock in" critical decision variables early on in the design process to keep the size of the search space tractable.

However, despite decades of experience, a heuristic-centric design philosophy can be overly restrictive (scanning a small part of the solution space) and lead to sub-optimal solutions. This is not only due to the highly non-linear nature of the search space but also because of the inherent complexity of handling multiple objective functions (emissions, economics, operational constraints etc.). Advances in computing and process systems engineering offer a route to systematically navigate the design space while accounting for critical design metrics and objectives.

This presentation will discuss the challenges in developing design tools for petrochemical separations focusing on distillation sequences. It builds on the pioneering work of Agrawal and co-workers (1-3) that relies on optimization models incorporating Underwood's shortcut method to minimize the vapor duty of a configuration. Furthermore, we discuss approximations that enable comparison of stream temperatures using only information about the composition, component relative volatilities and pressure ratios of the streams, thus making it possible to identify energy efficient distillation configurations.

The approach outlined here has been successfully applied to separation schemes of interest to the petrochemical industry resulting in configurations that not only provide energy savings (reducing emissions and operating costs) but are also operationally feasible. Moreover the capability to provide multiple options gives process developers the ability to adapt quickly to evolving design objectives. These results provide the impetus for further efforts in developing tools that effectively leverage both heuristics and advanced optimization, and usher in a new paradigm in chemical process design.

References

1. Agrawal, R., & Herron, D. M. (1997). Optimal thermodynamic feed conditions for distillation of ideal binary mixtures. AIChE journal, 43(11), 2984-2996.
2. Nallasivam, U., Shah, V. H., Shenvi, A. A., Tawarmalani, M., & Agrawal, R. (2013). Global optimization of multicomponent distillation configurations: 1. Need for a reliable global optimization algorithm. AIChE Journal, 59(3), 971-981.
3. Nallasivam, U., Shah, V. H., Shenvi, A. A., Huff, J., Tawarmalani, M., & Agrawal, R. (2016). Global optimization of multicomponent distillation configurations: 2. Enumeration based global minimization algorithm. AIChE Journal, 62(6), 2071-2086.

574722 Dynamic Simulation Studies for Optimal Design and Operation of an LNG Liquefaction Process

Mixed refrigeration cycle (MRC) natural gas liquefaction technology is wildly used for liquefied natural gas (LNG) production due to its unique advantages. Most of the research on LNG liquefaction process were based on steady-state simulation. Only a few addressed the dynamic performance of mixed refrigerant LNG processes. Thus, the main purpose of this study is to study the dynamic behaviors of mixed-refrigerant LNG liquefaction processes. The developed methodology in this paper contains three major tasks: steady state simulation model development, dynamic simulation with control structures to handle the disturbances and then disturbances were added to investigate behaviors of the process. In the first stage for a real LNG liquefaction process, a steady state simulation is conducted. In the second stage, the dynamic simulation which was based on the results of steady-state simulation was conducted. The results of steady-state simulation and dynamic simulation compared to keep the consistency. In the third stage, based on the dynamic model, various variations from natural gas temperature, pressure, flow rate, composition are introduced as disturbances to study the dynamic responses of the studied process. The dynamic responses of LNG temperature and compressor duty are the major performance evaluation criteria under these disturbances, from which the best control strategies are investigated to maintain the stable operation of the mixed-refrigerant liquefaction process. The simulation result showed that, compressor duty takes more time to stable than the LNG temperature. The variation of feed gas temperature has little influence on the process. The feed gas flow rate and methane composition have more effect than feed gas pressure and temperature effect. The feed gas composition has heat transfer performance and the flow rate affect. The feed gas pressure affects the process in two aspects: one is heat transfer performance and another one is the flow rate of natural gas.

555779 Crude Oil-to-Chemical Conceptual Design

Improving oil recovery technology has shifted attention away from concerns about peak oil supply[1]. With projections of increasing adoption of electric powered transportation vehicles, some oil producers now worry about peak oil demand. Depending upon your future view, oil demand will peak between ~2020 through 2050. At that point, countries with some of the largest oil reserves will still have ample available supplies. Full crude oil-to-chemicals (COTC) conversion is one approach to maintain demand for crude oil rather than abandoning it and losing the value of the remaining reserves.

CTOC produces non-fuels products from crude oil. Process routes involved can include combining existing refining and petrochemical units or developing new separation and conversion technologies.

Processing capacities will be large to meet economic targets imposed by economies of scale. The processing sequences involved require multiple steps. The combination will require large investments. Skillful conceptual design and layout of the overall CTOC has the potential for dramatic capital savings. Conceptual design of integrated CTOC complexes requires mastering three areas of knowledge:

- **Define the starting point.** What is crude oil? What are the variations in crude oil? What molecules are present?
- **Define the destination.** What are the products desired? What properties do they have? What shifts in product quality and composition can the market accept?
- **Define the path.** What are the routes from the feed to the product? What is the current technology status? What enabling processes are required? How can existing technologies and developing technologies be combined?

The presentation will examine these three areas and outline approaches that can be used in conceptual design.

[1] Hubbert's peak oil concepts.

561415 Design of a Single Wall Divided Wall Column for the Separation of an Industrial Quaternary Mixture

Distillation is a phase creation process involving addition and removal of heat. The distillation of tertiary and quaternary component mixtures is done using two and three fractionating columns respectively. The Divided wall column (DWC), is a single shell distillation column which is capable of separating mixtures of three or more components into high purity products, thus eliminating reboiler duties for the remaining columns. The separation of a tertiary mixture using the DWC is a known technology and has widely been used for the separation of various tertiary component mixtures. Separation of a quaternary mixture using a single wall Divided wall column was attempted. In this paper we have simulated a Divided wall column for the separation of styrene, benzene, ethylbenzene, toluene mixture. The simulation of the DWC was done using four column arrangement in Aspen Plus chemical process simulator. By using DWC arrangement we achieved a purity of 98.26% styrene, 97.47% ethylbenzene, 95.74% toluene and 99.99% benzene. The stage wise pressure, temperature and composition profiles along with the sizing and costing of the DWC are discussed in the paper.

565250 Process Intensification in the Methanol-to-Propylene Process Enabled By Rigorous Kinetics

In 1999 Lurgi, now part of Air Liquide, started the development of the so-called Lurgi MTP™ (methanol-to-propylene) process. This process is dedicated to the on-purpose production of propylene following a route to this valuable product that does not rely on crude oil as feedstock (as conventional propylene production technologies do), but instead utilizes natural gas or coal and potentially biomass. These alternative feedstock materials are first converted to synthesis gas, which is cleaned and then converted into methanol. After the first Lurgi MTP™ plant started up in 2010 and two more units followed in line (with a capacity of 470 kta propylene each), the MTP process is now an industrially referenced technology. Ongoing process optimization aims at improving the MTPprocess leading to higher yields, lower costs and generally CAPEX and OPEX savings.

In order to achieve substantial process intensification, it is essential to go into the details of the industrial process technology, particularly to clarify in a fundamental study the reaction kinetics of the MTP reaction network. . The reaction kinetic model then serves as a basis for reactor optimization using the innovative so-called Multi-Level-Reactor-Design methodology. In our work, more than 150 experiments describing 63 different reactions were performed in a Berty-reactor with dedicated analytics at FAU Erlangen to develop a complex kinetic model which has served as the basis for the study. The state of the art process was used as reference and the model was able to fully reproduce the main characteristics of the system. Test campaigns in a pilot plant at Air Liquide R&D in Frankfurt helped to validate the model. Figure 1 gives an example of investigating the optimum of productivity by variation of residence time using the kinetic model. In this presentation, we will demonstrate different examples to improve space time yield with modelling approaches and validation of computed results by pilot experiments. In conclusion, a significant potential for improving the process with new knowledge gained was identified, paving the way for process intensification in the MTP process.

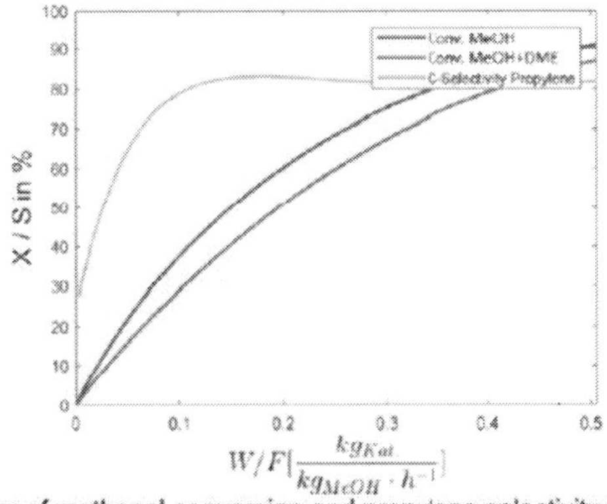

Figure 1 Dependency of methanol conversion and propylene selectivity on the length of reactor

- Abstract Text

566313 Oxidative Dehydrogenation for the Conversion of Ethane to Ethylene

The production of ethylene via steam pyrolysis represents a technology that is decades old and depending on the feed, may result in a wide range of byproducts, some of which may be valuable. Based on experimental yield data, process development and an associated economic analysis was performed. The basis of this analysis is a cost of production model that results in a required netback for the main product ethylene. This can then be assessed versus market pricing: contract or spot. These results were compared with the expected performance of a novel catalytic route based on the M1 catalyst. This catalyst, when running on ethane, demonstrates very high selectivity, while running at much lower temperatures and similar pressures to conventional pyrolysis. Results indicate that the new route has much lower variable cost, fixed costs and capital recovery possibly reaching shutdown economic level versus certain conventional processes. The yields and processing will be reviewed with an emphasis on the comparison to conventional technology.

574303 Techno-Economic Analysis of Direct Methane Aromatization for Natural Gas Upgrading with Mini-Plant Construction for Remote Deployment

Inexpensive sources of natural gas have led to a boom in development of technologies taking advantage of C1 chemistry; however, a significant number of potential natural gas sources are stranded and lack the infrastructure for economically-feasible transportation. The resulting gas, which is normally flared can, in theory, be recovered at the well-site, purified, and then processed to a product which is easier to transport. Gas-to-liquid (GTL) processes exist which can convert the feed to liquid products which can be more easily transported, although there are process constraints due to the need for deployment in remote locations, and several are being commercialized.[1] One such process, direct methane aromatization (DMA), produces significant amounts of benzene, a valuable commodity chemical, and hydrogen. The benzene can be purified on-site and then transported by truck; the hydrogen can be dealt with on-site as well by either transforming it to another transportable chemical, such as ammonia or cyclohexane, or using it to provide heating or electricity.

Previous economic studies of DMA have been reported which either invoke unrealistic assumptions in the process model,[2] do not consider processing of the gaseous hydrogen,[2, 3] or do not consider the limitations imposed on the process due to remote deployment.[4] With these issues in mind, a DMA mini-plant processing 20 MMSCFD of natural gas was designed with currently available technology to provide a platform for the investigation of process intensified alternatives. The reactor section of the plant uses a DMA catalyst developed at Texas Tech which deactivates at a significantly slower rate than previous catalysts.[5] The unit was investigated for economic feasibility and sustainability. In particular, the analysis identifies four particular areas which can benefit from intensification: aromatics yield (ie. one-pass conversion), coke selectivity, hydrogen recovery, and separation and purification of the aromatic products. Although intensified reactor technology for the DMA process is already being investigated for improving the yield with[6, 7] and without integrated hydrogen recovery[8, 9], the process model framework should provide opportunities for investigating additional intensification technologies, particularly in separations.

1. H. Fleisch. [White Paper], World Bank-Global Gas Flaring Reduction Partnership, 2015.
2. Pérez-Uresti, S.I. *et al.*, *Processes*, 2017, 5, 33.
3. Camilo Corredor, E., P. Chitta, and M. D. Deo. *Fuel Pro. Tech.*, 2019, 183, 55 – 61.
4. Huang, K. *et al.*, *Joule*, 2018, 2, 349 – 365.
5. Rahman, M., A. Sridhar, and S.J. Khatib. *Cat. A Gen.*, 2018, 558, 67 – 80.
6. Kee, B. *et al.*, *Eng. Chem. Res.*, 2017, 56, 3551 – 3559.
7. Morejundo, S.H. *et al.*, *Science*, 2016, 353(6299), 563 – 566.
8. Brady, C., B. Murphy, and B. Xu. *ACS Catal.*, 2017, 7(6), 3924 – 3928.
9. Cao, Z. *et al.*, *Chem. Int. Ed.*, 2013, 52, 13794 – 13797.

572855 Non-Oxidative Methane Conversion: Coupling Gas-Phase and Surface Models of Fe Single-Site Catalysts

Introduction

A better utilization of methane gas would lead to increased environmental and economic performance, particularly for resources that are currently flared. Non-oxidative conversion of methane (NOCM) has especially drawn considerable amount of interest for its simplicity and high selectivity. However, coke accumulation has been the biggest problem for NOCM.

A single site catalyst (SSC) may potentially resolve this issue by preventing adsorbate-adsorbate interaction for a coke growth with its sparsely located active sites [1]. In fact, Guo et al. [2] has shown a promising single site Fe catalyst on SiO_2 which showed stable catalytic activity without forming coke. Additionally, contribution from homogenous reactions to NOCM should not be overlooked because of a high NOCM temperature where homogeneous reactions actively occur. Therefore, we have combined SSC chemistry with gas-phase chemistry to model a NOCM process to comprehensively screen conditions for poly aromatic hydrocarbon (PAH) production which is indicative of coke formation.

Materials and Methods

Kinetic parameters of elementary steps for catalytic methane activation on SSC were calculated by *ab initio* density functional theory (DFT). A single Fe center with two neighboring carbons was modeled in a 2×2 cristobalite supercell. A Grid-Based Projector-Augmented Wave (GPAW) method with revised Perdew-Burke-Ernzerhof (RPBE) exchange-correlation functional was used in Atomic Simulation Environment (ASE). Barrier energies were calculated by finding a transition state with constrained minimization method.

For the gas-phase chemistry, the detailed kinetic mechanism was obtained from a recent literature study on methane pyrolysis to PAH formation [3]. Finally after combining surface chemistry species and gas-phase species into a single input file, Cantera was used to simulate constant pressure, isothermal batch reactor [4].

Results and Discussion

A complete free energy diagram with barrier energies for surface methane activation was obtained. The elementary reactions include methane dissociation, methyl radical desorption, and H_2 dissociation. The methane dissociation steps had the highest barriers.

Coupled heterogeneous and homogeneous reaction simulation was carried out with kinetic information calculated from barrier energies. The simulation was done for various reaction conditions including reaction time, pressure, inlet hydrogen concentration, and temperature to broadly illustrate different combinations of the controllable parameters. This result quantitatively suggests temperature and time windows for maximum ethylene production which is the most valuable product, and for minimum the coke precursor aromatics formation.

This work provides quantitative guidelines to design stable NOCM system which avoids coke formation. The inclusion of accurate gas-phase kinetics allows for systematic investigation of how the ideal catalyst could perform in a NOCM system.

References

1. Zhang, H., Liu, G., Shi, L., and Ye, J. *Energy Mater.* 8, 1701343 (2018).
2. Guo, X., Fang, G., Li, G., Ma, H., Fan, H., Yu, L., Ma, C., Wu, X., Deng, D., Wei, M., Tan, D., Si, R., Zhang, S., Li, J., Sun, L., Tang, Z., Pan, X., and Bao, X. *Science*. 344, 616 (2014).
3. Chu, T., Buras, Z.J., Oßwald, P., Liu, M., Goldman, M., and Green, W.H. *Chem. Chem. Phys.* In press.

4. Goodwin, D.G., Moffat, H.K., and Speth, R.L. *Cantera: An object-oriented software toolkit for chemical kinetics, thermodynamics, and transport processes.* https://www.cantera.org, 2018. Version 2.4.0. doi:10.5281/zenodo.170284

567956 Catalyst Design for Microwave Assisted Methane Dehydroaromatization

The abundance of domestic natural gas reserves has resulted in increasing interests in generating higher value chemicals from methane, the main component of natural gas. While methane upgrading is currently achieved through an indirect route via syngas, direct upgrading offers conceptually a less energy intensive and costly alternative. One promising direct upgrading route is the non-oxidative conversion of methane to aromatics via dehydroaromatization (DHA). However, at the very high temperatures thermodynamically required for this reaction to proceed, carbon deposition becomes inevitable and results in rapid deactivation of the catalyst, rendering the process uneconomic. In the present project, we are exploring a novel process for methane upgrading by applying microwave radiation. Microwave radiation could provide a much more efficient way than conventional heating to couple energy into the reaction system, and is expected to do so more selectively by selectively heating the active (metal) site of the catalyst while leaving the bulk catalyst and gas phase at a lower temperature to avoid or reduce carbon formation. However, due to their complexity, microwave-assisted heterogeneous reactions are poorly understood to-date and systematic studies on how microwaves interact with catalytic materials are largely lacking. Here, we present results from an on-going study that aims to improve the understanding of catalyst design for microwave catalysis by systematically exploring and identifying catalyst properties that affect microwave absorption.

The catalyst for methane DHA is typically composed of metal nanoparticles dispersed in zeolites. In these catalyst systems, parameters including zeolite particle size, the concentration of Brønsted acid sites, speciation and dispersion of metals, and the nature of the metal can be expected to not only affect the catalytic reactivity but also to influence the ability of the catalyst to absorb and interact with microwave radiation. The main objective of our study is hence to obtain an improved understanding of what gives metal/zeolite-based catalyst higher microwave sensitivity. Towards this goal, we synthesized carefully defined metal/ZSM-5 catalysts to individually vary the size and Si/Al ratio of the zeolite ZSM-5, as well as speciation and type of metal in the catalyst. We find that incorporating more Al into the ZSM-5 framework increases the microwave sensitivity, while the size of ZSM-5 has a comparatively small effect. On the metal side, Fe-ZSM-5 was prepared via three different synthetic approaches to control metal dispersion and speciation. We find that both metal speciation and metal loading have significant impact on microwave sensitivity. Further studies regarding the nature of the metal are currently on-going. The observed effects will be presented and discussed in detail in the presentation.

Overall, through these studies, we aim to develop a fundamental understanding of metal/zeolite catalysts in microwave catalysis as a basis for rational catalyst design for microwave-assisted methane DHA.

573510 Microwave Assisted Catalytic Conversion of Ethane to Aromatics for a More Efficient Approach over a Conventional Fixed Bed Reactor

The mechanism of ethane dehydroaromatization over a 4%Mo-0.5%Fe/ZSM-5 catalyst was tested in a 6.650 GHz microwave fixed bed reactor (MW) and compared to a conventional fixed bed reactor (CFB). Over 80% ethane conversion was achieved under microwave irradiation at 375°C bulk temperature. The conventional fixed bed required a corresponding temperature of 660°C to achieve the same conversion as the microwave fixed bed reactor. The selectivity's of the catalyst were compared after 10 minutes of time-on-stream where the conversion of ethane was the same and reactant stream concentrations were fully developed. The CFB had a higher selectivity to aromatics; however, the MW presented a selectivity to ethylene of 26.5% higher than the CFB. The MW high selectivity to ethylene suggest that the aromatization reaction was limited possibly do to ethylene's inability to rapidly diffusion into the pores of the zeolite structure. This was attributed to the overall low bulk temperature of the catalyst bed or uneven heat distribution throughout the catalyst bed. TPO analysis of CFB showed three distinct orders of coke formation, whereas the microwave resulted in a single higher order temperature burring coke. Less total coke formation was observed on the MW spent catalyst, this was attributed to the low bulk temperature in the microwave which suppressed the formation of amorphous coke on the catalyst surface.

572117 Effects of Low Dopant Concentrations on the Performance of Samarium Oxide in the Oxidative Coupling of Methane

The oxidative coupling of methane (OCM) is a potential synthesis route to produce ethylene directly from methane, or natural gas, in one step. A number of different materials have been investigated in this reaction, but few catalysts display high activity, thermally stability and selectivity towards C_{2+} products (ethane, ethylene and higher hydrocarbons). Some of the most promising catalysts tend to be complex mixtures of oxides which makes identifying structure activity relationships challenging. To simplify the structure activity relationship in order to investigate fundamental effects, we have chosen to study the effects of low molar dopant concentrations on samarium oxide (Sm_2O_3) in the oxidative coupling on methane. We found that even at concentrations as low as 0.1 and 1.0 mol %, the dopants affect the methane conversion and the selectivity to ethane and ethylene of samaria with the highest increases in catalyst efficacy observed on 0.1% loading. Catalyst characterizations reveal that the dopants affect the crystal structure, as well as the number and strength of the basic sites, of samaria both of which are expected to influence the OCM properties.

569296 Enhancing Selectivity in Oxidative Coupling of Methane Using Catalytic Solid Oxide Membrane Reacting Systems

Oxidative coupling of methane (OCM) is a chemical process in which methane is converted directly into ethylene and ethane (C_2) at high temperatures (600-900°C) in the presence of oxygen and typically an active catalyst. The main challenge with this process is the formation of thermodynamically stable over-oxidation products, mainly CO and CO_2, which limits the C_2 selectivity. This selectivity problem is particularly significant in conventional reactors (co-fed with methane and oxygen) due to the presence of large gas-phase oxygen concentrations at the reactor inlet, which accelerates the unselective, deep oxidation reactions.[1] Catalytic solid oxide membrane reactors can minimize the unselective reactions by separating the methane and oxygen feed, and allowing the selective passage (conduction) of oxygen ions (O^{2-}) from the oxygen side to the methane side where the reaction occurs.[2,3] By allowing the distributed and controlled delivery of O^{2-} to the active sites, gas-phase oxygen can be significantly minimized, thereby limiting the occurrence of deep oxidation reactions. These membrane systems offer additional benefits including better control of reactor temperature and simultaneous separation of pure oxygen from other components in air thereby simplifying downstream separation. Herein, we have synthesized and tested catalytic solid oxide membrane systems in OCM. We show that a $BaCe_{0.8}Gd_{0.2}O_{3-\delta}$ (BCG) membrane system exhibits stable OCM performance and high carbon tolerance at extremely low oxidizing conditions. We also demonstrate that the BCG membrane system operated in membrane mode (separated and distributed oxygen feed) gives significantly higher C_2 selectivity compared to the same system co-fed with methane and oxygen, at comparable methane conversions. These results provide reliable experimental evidence of the improved OCM performance of a catalytic membrane system over a conventional co-fed system.

References

(1) Farrell, B. L.; Igenegbai, V. O.; Linic, S. *ACS Catal.* **2016**, *6* (7), 4340–4346.

(2) Farrell, B. L.; Linic, S. *Catal. Sci. Technol.* **2016**, *6*, 4370–4376.

(3) Igenegbai, V. O.; Meyer, R. J.; Linic, S. *Appl. Catal. B Environ.* **2018**, *230*, 29–35.

559199 Multi-Scale Simulation Framework for the Oxidative Coupling of Methane: From Process Layouts to Catalyst Particles

Introduction

The Oxidative Coupling of Methane (OCM) represents a direct valorization route for natural/shale/bio gas resources which has captivated both academic and industrial stakeholders for almost 40 years[1]. It is widely recognized that the main challenges of OCM, situated in difficult CH_4 activation combined with an inverse relationship between methane conversion and selectivity towards the desired C_2 (C_2H_6, C_2H_4) products and strong exothermicity, can be overcome only via simultaneous optimization of process layout, reactor design and catalyst composition[2]. In fact, the only claimed commercial implementation so far, the GEMINI process by Siluria and Linde[3], has introduced significant novelties in all these three fields.

Nevertheless, most of the OCM process studies currently available in literature are based on a single catalyst and a single reactor configuration. The performance indicators at the outlet of the OCM reactor are 'imposed'[4] or simulated with simplified kinetics[5]. Variations in CH_4 conversion and C_2 selectivity with the process conditions or catalyst composition would have a dramatic impact on the separation and recycle sections downstream and cannot be easily extrapolated to other catalysts.

In order to overcome these limitations, in the present work a multi-scale simulation framework is proposed which enables process modelling for OCM by incorporating a detailed reactor model and a fundamental kinetic model specifically accounting for the catalyst properties.

Simulation framework

A schematic representation of the simulation framework developed is shown in Figure 1. Three different modelling challenges were tackled and are displayed in the figure:

Process model

Process simulations were performed via the commercial software Aspen Plus. An example of a potential process flowsheet[6] is shown in Figure 1/A. The major challenge of the present work lies in the detailed modelling of the reaction section. This was tackled by coupling Aspen Plus to an external, in-house code. The implementation occurred via a customized user unit operation model[7] (Figure 1/B1), namely a Fortran interface which is capable of: reading input from Aspen streams and block variables, write input to the external code, execute the external code, read output from the external code, write output to Aspen streams.

Reactor model

The external code solves mass, energy and momentum balances for a catalytic fixed-bed reactor (Figure 1/B1), which can be operated both isothermally and adiabatically[8]. The reactor model is 1-dimensional, i.e. no radial gradients are considered on the reactor scale, and heterogeneous, i.e. irreducible concentration and temperature gradients on the particle scale are taken into account. The two phases considered on the particle scale, and indicated in Figure 1/B2, are: the *intraparticle* phase, consisting of the catalyst particles and the gas contained in their pores, and the *interstitial* phase, accounting for the gas around the particles. Textural properties of the catalyst particles (surface area, porosity, tortuosity, diameter) are included among the model parameters.

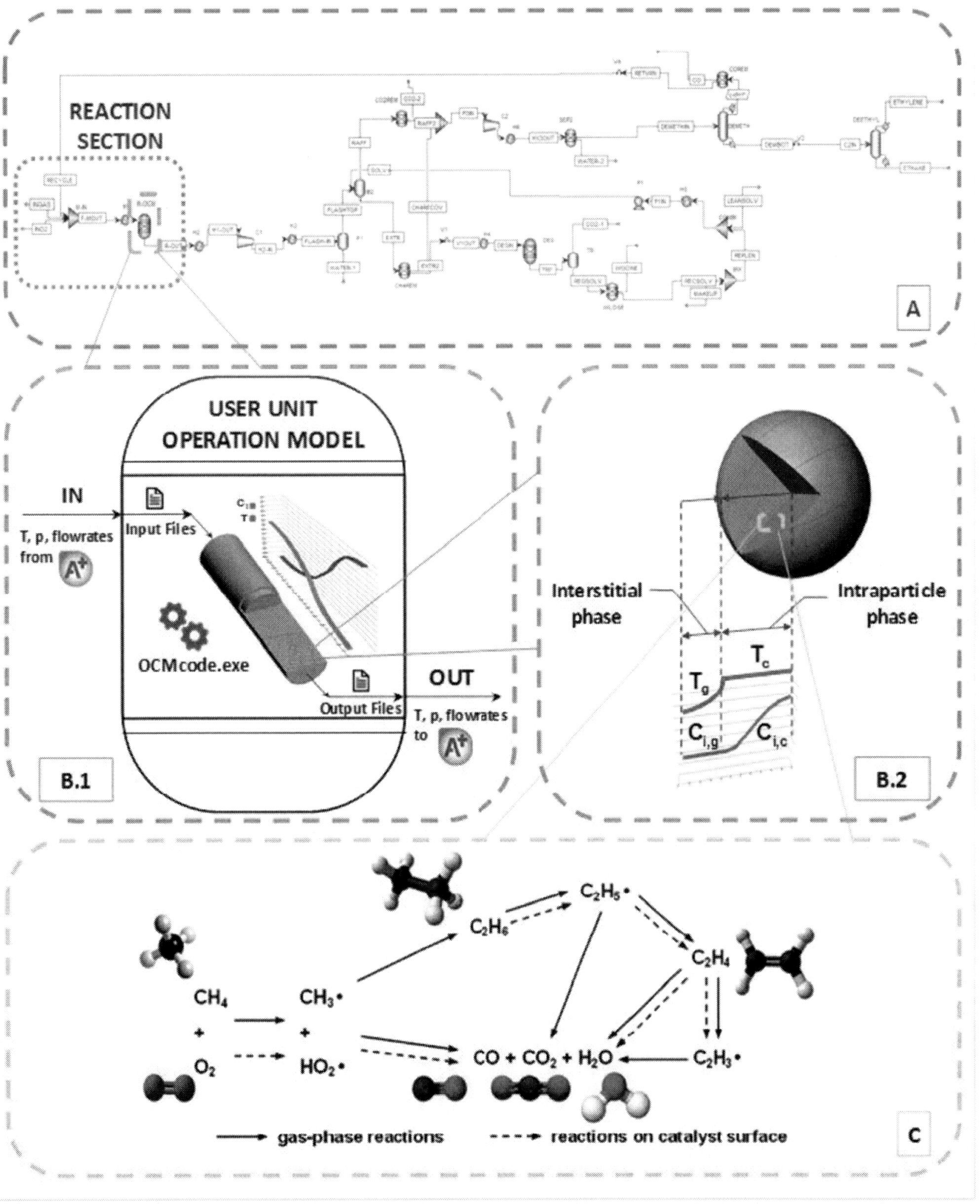

Figure 1. Multi-scale simulation framework deployed in the present work. A. Example of process flowsheet for OCM[6]; B. Customized user unit operation model, enabling the simulation of an OCM reactor (B.1), including particle-scale phenomena (B.2)[8]; C. Simplified representation of the reaction scheme[9].

Microkinetic model

A microkinetic model for the OCM reaction, previously developed in our research group and validated for five different catalysts[9], was embedded in the reactor model. The model accounts for 13 molecules, 11 radicals and 11 surface species, and comprises 39 reversible gas-phase reactions and 26 reversible reactions on the catalyst surface. A simplified representation of the reaction scheme is reported in Figure 1/C. The possibility of reproducing the performances of different catalysts arises from the presence in the model of catalyst descriptors, i.e. kinetically significant catalyst features - such as chemisorption enthalpies, sticking coefficients, density of active sites - that specifically account for the impact of the catalytic material on the reaction kinetics.

Case study: distributed oxygen feed and intermediate cooling

In this case study, the reaction section of a potential OCM plant was simulated considering a Sn-Li/MgO catalyst[9]. The base case, corresponding to scenario A.I in Figure 2, was characterized by a single fixed-bed reactor operated adiabatically, at atmospheric pressure, with a methane-to-oxygen molar ratio equal to 10. The space time was set to 17 kg s/mol$_{CH4,0}$ via a design specification to achieve complete oxygen conversion at the reactor outlet. The inlet temperature was chosen equal to 840K via a design specification to operate the reactor with maximum temperature equal to 1273K[8]. The results obtained in terms of CH_4 conversion, C_2 selectivity and C_2 yield are reported in the first column of Table 1.

A second scenario, indicated with A.II, was simulated by considering the same process conditions as A.I, but reducing CH_4/O_2 to 5. As highlighted in the second column of Table 1, the resulting outlet temperature was way beyond the limit of 1273K, due to the higher adiabatic temperature rise associated to the higher oxygen content. This scenario exemplifies that not all sets of process conditions are possible for the operation of an adiabatic OCM reactor[8], and, hence, was discarded.

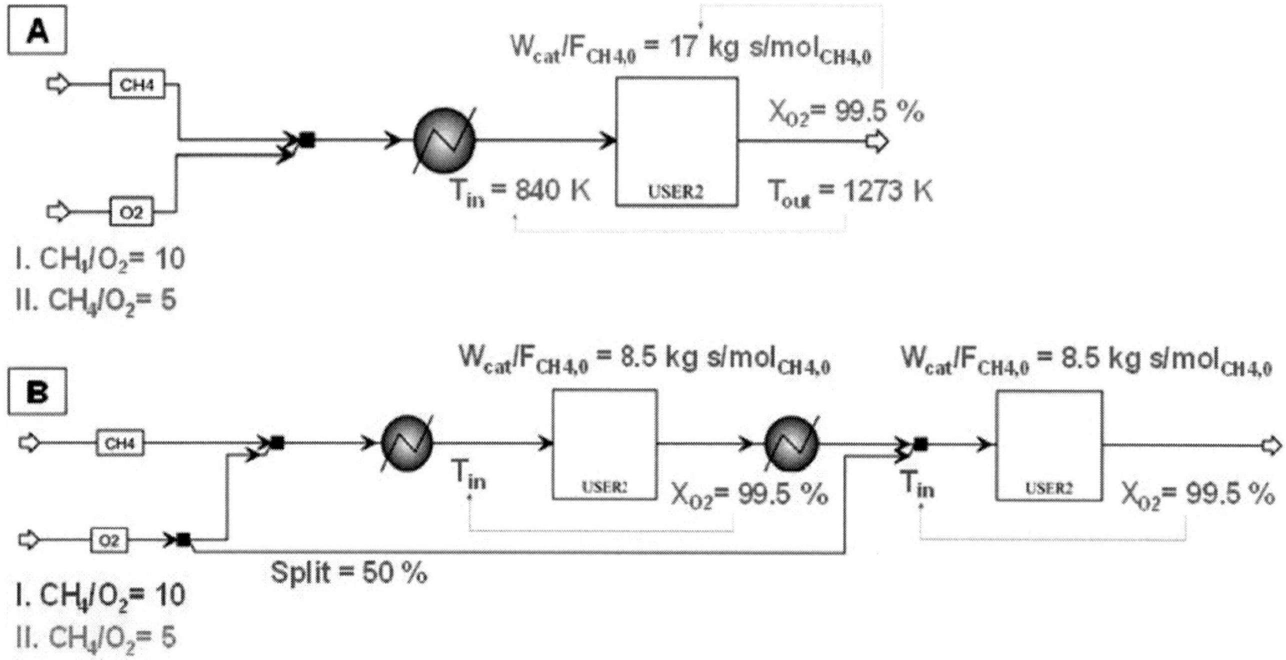

Figure 2. Process layout of the OCM reaction section in the present case study, based on Sn-Li/MgO catalyst. The values of CH_4/O_2 reported are referred to the global methane and oxygen molar flowrates.

Scenario B.I was characterized by two Sn-Li/MgO catalyst beds in series, with distributed oxygen and intermediate cooling. The amounts of CH_4, O_2 and catalyst were kept constant in comparison with scenario A.I, but catalyst and oxygen were split in two, with 50% of the oxygen being fed to the first bed, containing half of the catalyst, while the other 50% was fed to the second bed. The inlet temperature to each bed was adapted via a design specification, again to achieve complete oxygen conversion at the reactor outlet. As it can be observed in the corresponding column in Table 1, this scenario resulted in higher inlet temperatures. This is due the fact that catalytic light-off is more difficult to achieve when methane is present in high excess[8]. The corresponding outlet temperatures were lower for both beds, due to the limited methane conversion that was obtained with such a low oxygen content. This resulted in a lower C_2 yield as compared to the base case.

Scenario B.II was analogous to B.I, this time considering $CH_4/O_2 = 5$. In Table 1 can be observed that, compared to the base case, this process configuration resulted in an increase of +22% in the yield. The price to pay was +40K in the required inlet temperature and -19% decrease in the selectivity.

As the computational time for each simulation was in the order of 30s, the present case study could easily serve as basis for further techno-economic analyses aimed at evaluating the profitability of different configurations, such as scenarios A.I (higher selectivity) and B.II (higher yield), embedded in a full process scheme, including also the separation and the purification sections and the recycle of material streams.

Table 1. Simulation results for the present case study. For all the scenarios: p= 1bar, $W_{cat}/F_{CH4,0}$= 17 kg_{cat} s/mol_{CH4}. The variables of major interest are indicated in bold.

	Scenario A		Scenario B (multiple beds with intermediate cooling)					
	I	II	I			II		
N. of beds	1	1	2			2		
			Global	First Bed	Second Bed	Global	First Bed	Second Bed
CH_4/O_2	10	5	10	20	18	5	10	9
T_{in}	840K (manipulated)	840K	-	883K (manipulated)	861K (manipulated)	-	880K (manipulated)	855K (manipulated)
T_{out}	1273K (controlled)	1425K		1125K	1111K	-	1273K	1263K
X_{O2}	100% (controlled)	100%	100%	100% (controlled)	100% (controlled)	100%	100% (controlled)	100% (controlled)
X_{CH4}	15%	31%	12%	8%	4%	22%	14%	20%
S_{C2}	75%	64%	66%	77%	46%	63%	76%	40%
Y_{C2}	11%	20%	8%	6%	2%	14%	11%	4%

The few variations investigated in this case study were mainly aimed at demonstrating the potential of the developed simulation framework and more scenarios are currently being pursued. For instance, ongoing simulations evaluate how activities and selectivities of alternative catalysts, such as Sr/La_2O_3 and $NaMnW/SiO_2$, impact the optimal process configuration identified from yield and/or selectivity maximization.

Conclusions

The simulation framework developed in the present work enables multi-scale modelling of the OCM process, without compromising on transport phenomena, both on reactor and particle scale, and kinetics, which is herein described in detail via a microkinetic model. The enablement of process simulations based on microkinetic models represents a step forward in the design and assessment of realistic process layouts for natural gas valorization. In fact, the possibility of evaluating the impact of process conditions, feed composition and catalyst properties on the outlet stream from the reaction section paves the way for more reliable techno-economic assessments aimed at designing an industrially-viable OCM process.

References

1. Keller, G.E. and M.M. Bhasin, Journal of Catalysis, 1982

2. Kondratenko, E.V., et al., Catalysis Science & Technology, 2017

3. https://www.linde-engineering.com/en/process_plants/petrochemical-plants/oxidative-coupling-methane/index.html

4. Ortiz-Espinoza, A.P., et al., Computers & Chemical Engineering, 2017

5. Godini, H.R., et al., Fuel Processing Technology, 2013

6. Arickx, M., et al., Internal Report, 2018

7. Tripodi, A., et al., Catalysts, 2017

8. Pirro, L., et al., Industrial & Engineering Chemistry Research, 2018

9. Alexiadis, V.I., et al., Applied Catalysis B, 2016

560651 Single Site Vs Crystalline: Development and Characterization of Mn/W/Na/SiO2 Model Catalysts for the Oxidative Coupling of Methane (OCM)

An abundance of natural gas due to the advancements in exploration and extraction technologies has rejuvenated interest in the direct conversion of natural gas to ethylene via oxidative coupling of methane (OCM). The tri-metal oxide supported Mn/W/Na/SiO2 catalyst has been studied in the past and has shown ~75% C_2 selectivity, ~35% conversion and robust thermal stability under operating conditions. However, improvements need to be made to the catalyst to increase product yield and lower the operating temperature. Most published studies of the tri-metal oxide catalysts before and after OCM typically identify crystalline phases such as Na2WO4, MnWO4, etc. as the active phases. Our in-situ and operando Raman spectroscopy studies of supported Mn/W/Na/SiO2 catalysts during OCM, surprisingly, reveal that the often suggested crystalline phases are not present during reaction and that dispersed surface metal oxide species are present. This suggests that the surface metal oxide species may be the catalytic active sites for OCM by supported Mn/W/Na/SiO_2 catalysts. Developments regarding design and performance of dispersed and crystalline tri-metal oxide catalysts for OCM at/near working conditions will be presented in this work. The focus will be on *in-situ*, and *operando* Raman spectroscopic characterization of the model tri-metal oxide Mn/W/Na/SiO_2 catalysts to establish the nature of the surface active sites, their H_2-TPR characteristics, and their structure dependent catalytic performance.

579721 Optimization of C3MR LNG Liquefaction Process Using a Parallelized Simulator

Controlling the cost of energy-intensive processes such as C3MR (propane/mixed refrigerant) LNG liquefaction is a prime opportunity for optimization. Our work combines different C3MR models to develop an optimization that uses multiple independent variables, relaxed constraints, and flowsheet alterations to achieve an optimal operating cost. This work uses a rigorous model of the process in a process simulator rather than a reduced model of the process response. Use of a first-principles process simulator increases the accuracy of the model and therefore allows a more realistic optimization, at the cost of increased computing time. Parallelization (using multiple CPU cores) in the process simulator greatly increases the speed of the optimization calculations. CHEMCAD's optimization tools and multi-core technology make it possible to easily define the optimizations, run the optimizations efficiently, and generate global optima with a high degree of confidence.

569014 Study on Advanced Control for Natural Gas Dehydration Plant

Mono ethylene glycol (MEG) based natural gas dehydration process is facing significant glycol losses through stripper column due to vaporization. The regulatory control strategy can minimize this problem at a certain level. This strategy can perform effectively under certain upset conditions as well [1]. However, to run the plant more efficiently and economically it is necessary to apply advanced control technologies to absorb more severe process upsets and eliminate MEG losses as much as possible. In this study, a plant-wide dynamic simulation model for a natural gas dehydration plant has been firstly developed and examined. Based on that, a hierarchical control system comprised of Dynamic Matrix Control (DMC) and basic regulatory control loops is constructed to optimize the plant operation in terms eliminating MEG losses and minimize the operating costs with ensured product qualities under various process upsets. Rigorous DMCplus and DMC3 (adaptative dynamic matrix controller) controller models are respectively developed based on the subspace identification model and simulated with Aspen manufacturing platform. Both DMCplus and DMC3 controllers show promising results of minimizing MEG losses by over 90%. Although DMC3 perform a little better than DMCplus, both controllers are actually very robust, which can handle ±25% feed and ±25% chiller temperature disturbances. Through virtual simulations, this study also shows that the developed hierarchical control system outperforms the regulatory PID controller strategy in minimizing MEG losses under process constraints.

References:

1. Haque, M.E.; Xu, Q.; Palanki, S. Glycol Loss Minimization for a Natural Gas Dehydration Plant under Upset Conditions, *Eng. Chem. Res.* **2019**, 58, 1994-2008.

573187 Novel Piecewise Linearization of Rigorous Nonlinear Natural Gas Transportation Model for Optimal Network Operation

Natural gas has become a vital energy source due to its electricity generation efficiency and low emissions relative to other fossil fuels. Currently natural gas accounts for an estimated 22% of the total global energy mix, and conservative estimates in sustainable energy growth predicts the share of natural gas to grow to 25% of the global energy mix by 2040 [1]. The large and growing demand for natural gas makes the operational challenges associated with its transportation ever-important. Operations research has the potential to significantly improve the efficient operation of natural gas transportation systems if accurate solutions can be obtained within practical time constraints. For instance, an estimated 3-5% of all natural gas transported by pipeline is consumed by the compressor stations responsible for maintaining gas flow. It has been estimated that at least a 20% reduction in compressor station fuel consumption can be obtained through global optimization of pipeline operation [2]. However, natural gas transportation system operation is highly dynamic due to frequently changing uncontrollable variables, such as customer demand and available gas supply. As operating conditions change, the optimal operating point also changes, so solution must be attainable relatively rapidly for operations research to be useful for guiding the online operation of natural gas transportation systems.

Modelling the operation of natural gas transportation through pipelines in a rigorous way requires the use of a large set of nonlinear equations to capture gas flow and compressor station physics [3]. Further, integer variables are often required to model equipment states and bidirectional gas flow [4]. As a result, natural gas transportation is most rigorously modelled as a mixed-integer nonlinear program (MINLP). The computational complexity of this problem class means that applying rigorous operations research methods to guide natural gas transportation operation has been limited to small problems which are not representative of those faced in industry.

To enable solution of large-scale problems major simplifications to the rigorous MINLP model have been employed. A common simplification strategy is to assume certain modelling parameters such as the friction factor, gas compressibility factor, gas isentropic exponent, and compressor efficiency to be constants, and their nonlinear relationships to the system operating conditions are ignored. The drawback of this approach is that the error introduced into the model through these simplifications is significant, often resulting in solutions which are found to be infeasible when tested on a rigorous simulation [5].

There are also very few operations research studies that consider the ability for pipelines to act as storage vessels, through what is referred to as linepacking. This capability of pipelines is a valuable tool that can be used as a buffer to mitigate the interaction between variable upstream and downstream operating conditions, enabling both a better security of gas supply and the ability to take advantage of gas market fluctuations [6]. In large gas transportation networks the storage capacity available through linepacking is significant, so the network operators heavily consider linepacking in how they operate the network. Determining an optimal linepacking policy requires accurate modelling of compressor stations and incorporation future operating uncertainty in the transportation model. This additional computation complexity has resulted in the vast majority of operations research to-date ignoring the capability of pipelines to store linepack, and focus simply on the compressor fuel cost minimization problem [6].

By using a model of natural gas transportation that ignores the role of linepacking in gas network operation, and employs the common simplifications found in the literature, the practical usefulness of optimal solutions obtained is severely limited. This work aims to increase the practical usefulness of operations research in guiding gas network operation by increasing the modelling accuracy over the traditional approaches, while ensuring the model can be solved on large-scale networks rapidly. To do so, a novel linearization approach is used to capture the complex nonlinear relationships that are often ignored in the simplified models. The linearization method is an MILP that produces the optimal piecewise-linear function which most closely approximates the relationship being approximated. In our previous work we introduced the optimal piecewise-linear function generation method to approximate the relationships of the gas compressibility factor, gas isentropic exponent, and compressor efficiency with the pipeline operating condition [5]. We demonstrated that a model using the piecewise-linear function approximation obtains optimal solutions that have less than 1% error when compared to a rigorous simulation. Further, the solution time of the novel partially piecewise-linear model is shown to be at least an order of magnitude less than the typical MINLP model considered as the most rigorous [5].

In the previous work the nonlinearity in the gas transportation model was only partially approximated by piecewise-linear functions, and line-packing was not considered. This work fully linearizes the gas transportation model using the same novel piecewise-linear function generating technique employed previously. Additionally, a novel linepack modelling formulation is introduced which uses a two-stage stochastic modelling framework to capture uncertainty in future pipeline operating conditions. A case-study is performed which demonstrates that despite the drastically increased problem size due to the consideration of line-pack, optimal solutions to a large-scale gas network representative of the Norwegian pipeline system are able to be obtained within 1 minute when using the novel linearized model. Further, the optimal solutions from the novel linearized model are compared to a rigorous simulation which demonstrates that they contain less than 1% error. For comparison, the typical MINLP model was also applied in the case study and was only able to find a feasible solution after 12 hours.

The case study results demonstrate that the novel linearized model can substantially reduce solution time and maintain the solution accuracy compared to the typical MINLP model that is currently consider as the most rigorous natural gas optimization model available. Overall, by applying the novel linearization technique to approximate the nonlinear relationships inherent in natural gas transportation modelling the model complexity is drastically reduced, and the model accuracy maintained. The significance of these results is that accurate solutions to problems of practical importance can be obtained within reasonable solution times, thereby improving the usefulness of operations research in guiding online operation of natural gas transportation systems.

References

[1] IEA. *World Energy Outlook 2018*. International Energy Agency, Paris, **2018.**

[2] Schroeder, D. Hydraulic analysis in the natural gas industry. In *Advances in Industrial Engineering Applications and Practice I*; International Journal of Industrial Engineering: Houston, TX, USA, **1996**, 960–965.

[3] Schmidt, M.; Steinbach, M.C.; Willert, B.M. High detail stationary optimization models for gas networks. *Optim. Eng.* **2015**, *16*, 131–164.

[4] Wu, S.; Rios-Mercado, R.; Boyd, E.; Scott, L. Model Relaxations for the Fuel Cost Minimization of Steady-State Gas Pipeline Networks. *Math. Comput. Model.* **2000**, *31*, 197–220.

[5] Kody Kazda and Xiang Li. Approximating nonlinear relationships for optimal operation of natural gas transport networks. *Processes,* **2018,** *6(10),* 198.

[6] Roger Z. Rios-Mercado and Conrado Borraz-Sanchez. Optimization problems in natural gas transportation systems: A state-of-the-art review. *Applied Energy,* **2015**, *147*, 536-555.

573242 Simulation of Pressure Swing Adsorbers Using the Moving Bed Model

Simulation of pressure swing adsorption (PSA) units remains a challenge for practicing engineers, which hinders the process intensification potential of the technology. The Moving Bed model, a steady-state model which represents any adsorption process as a series of interconnected continuous adsorption units, has the potential to alleviate some of the difficulties in modeling PSA.[1, 2, 3] In the Moving Bed model, the cycle is divided into multiple constant-pressure operating states bearing similarity to the individual steps of the cycle which allows the properties of the state to be evaluated from equilibrium considerations alone. The operating steps are arranged from highest-to-lowest pressure to represent the driving force gradient exploited during the cycle and then connected by a virtual moving bed. The virtual moving bed does not physically exist in the system but represents the adsorbate and void space hold-up retained in the adsorption column at the end of the operating state. However, modeling a cyclic batch process as a continuous, equilibrium process neglects the rate-based transport phenomena involved. Using the idea of adsorption efficiencies[4] to account for non-idealities like mass transfer resistance and dispersion, similar to the tray efficiencies used in rate-based distillation modeling, the retentate flow rates can be adjusted to match the actual performance of the unit.[2, 3] To showcase the capabilities and current limitations of the model, a case study will be presented which investigates multiple PSA units undergoing different types of separations with different adsorbents and cycles. Guidelines for ensuring convergence and robustness will be given and the effect of important parameters will be discussed.

1. Sees, M.D. (2018, October). *Novel Steady-State Process Modeling Methodology for Pressure Swing Adsorption.* Paper presented at the 2018 Annual Meeting of the American Institute of Chemical Engineers, Pittsburgh, PA.
2. Kirkes, T. (2018, October). *Steady-State Modeling of Air Separation by Pressure Swing Adsorption Using a Bed-State Efficiency Model.* Poster presented at the 2018 Annual Meeting of the American Institute of Chemical Engineers, Pittsburgh, PA.
3. Sees, M.D. (2019, April). *Modeling of Gas Separations by Pressure Swing Adsorption Using a Novel Steady-State Methodology.* Paper presented at the 2019 Spring Meeting of the American Institute of Chemical Engineers, New Orleans, LA.
4. Smith, O.J. and A.W. Westerberg. *Eng. Sci.*, 1991, 48(12), 2967 – 2976.

563191 Correlation between the Yield of Deasphalted Oil and Hansen Solubility Parameter of Its Feedstock in a Solvent Deasphalting Process

For using heavy oil energy source, the asphaltene components leading to coke formation and catalyst poisoning should be eliminated. Among the heavy oil upgrading processes, solvent deasphalting (SDA) process is specialized in the selective removal of asphaltene components. The feedstock of SDA process is substantially viscous and high-molecular-weight oil such as vacuum residue and bitumen, and C3-C6 alkane is used as the extraction solvent. The extraction solvent selectively dissolves deasphalted oil and rejects asphaltene components that are not dissolved into the solvent.

It is expected that the modeling of the SDA extraction can be very effective for the development of the process though, the study of the modeling of the SDA extraction is at the basic level since the structure and phase equilibrium of heavy oil is still controversy. In the previous study, only single solubility parameter difference between the feedstock and extraction solvent was used as roughly verifying the relationship between the yield of deasphalted oil. However, in this study, it was examined that the relationship between the yield of deasphalted oil and relative energy difference, which was calculated from (1) three Hansen solubility parameters of the feedstock and extraction solvent, and (2) extraction conditions such as solvent types, temperature, and solvent to oil ratio. Incidentally, the feedstock was separated into SARA components and their Hansen solubility parameter were obtained because the solubility levels of heavy oil feedstock range widely.

574173 New Digital Method Enables Flexibility, Production, and Reliability Improvements While Mitigating Overhead Corrosion from Tramp Amines

In the pursuit of efficient and profitable refinery operations, owner-operators must continuously balance feedstock flexibility and product optimization with both mechanical availability and long-/short-term asset-protection. Further complicating this process is the limited ability to effectively determine which operational adjustments are most impactful. Many refineries today sometimes face periods of severe corrosion and associated fouling in the internals of atmospheric towers and overhead condensate systems. Often the corrosion problems are tied to incoming crude changes and considered largely unavoidable. This situation can lead to implementation of overly conservative short or long-term solutions at the exclusion of more complex but effective ones. Over time, the lack of ability to rapidly detect, define, analyze and respond to the rapidly changing drivers of corrosion has dramatically limited the effectiveness of practical solutions. This situation unnecessarily increases the costs and operational risks at a refinery.

Recent introductions of equipment allowing real-time monitoring of key overhead parameters has helped reveal a variety of continuing episodic system fluctuations leading to significant impact on corrosion. In parallel, high frequency and low latency amine speciation analysis is bringing ongoing information about crude contaminant changes and their impacts on the overhead system. To take advantage of the wealth of information available today, new digital platforms capable of effectively handling the data are needed. By increasing the frequency and sophistication with which salt formation data are collected and analyzed, refineries can achieve more proactive ways of controlling corrosion and fouling while simultaneously achieving production and quality targets. This paper will present real examples taken from operating refineries to illustrate new and effective ways of mitigating overhead corrosion problems using streaming telemetry and sophisticated data analytics.

575652 Study on Wave Motion Effects on FLNG Mixed Refrigeration Process Via Integrated CFD and Process Simulations

Floating liquefied natural gas facility (FLNG) has been studied since 1970, but makes very low progress until recent years, Compared with the traditional liquefaction facilities, FLNG faces several challenges, such as it has only very limited space to contain many necessary units; meanwhile, sea wave motions will bring negative impacts on the stability and safety of the onboard manufacturing process. During the FLNG liquefaction process, the mixed refrigerant (MR) separator would easily be affected by the wave motions, so that the separated vapor stream would have some liquid entrainment, which would further degrade the MR refrigeration performance. Hitherto, there are few researches on how sea wave motions influence the MR process.

In this study, the dynamic effects of process under wave motion was investigated by utilizing both computational fluid dynamics (CFD) and process dynamic simulation, a commercial CFD tool was applied and a multiphase model was developed to simulate the fluid moving and separation efficiency within the MR separator. A user-defined function (UDF) has been developed in C and hooked with the solver to model the wave motion and liquid sloshing. Based on the CFD simulation, MR separator efficiency data has been obtained and coupled with a process simulation model to quantitatively study the impact of refrigeration system performance. This study could help future studies on process optimization to improve process design, process controllability, and safety of the FLNG liquefaction process.

569806 Understanding the Role of Partial Oxidation Reactions during Autothermal Pyrolysis

The goal of this project is to understand the partial oxidation reactions responsible for energy release during biomass autothermal pyrolysis. The enthalpy for pyrolysis is provided by exothermic oxidation reactions within the reactor instead of external energy supplied by heat transfer. This is achieved by reacting a small amount of oxygen (equivalence ratios of 0.06-0.10) with biomass in a fluidized bed reactor, which partially combusts the products of pyrolysis to overcome the endothermic enthalpy of pyrolysis. Experiments performed at Iowa State have shown that autothermal pyrolysis does not produce significant loss in bio-oil yield or quality as compared to traditional non-oxidative pyrolysis. Consequently, we are developing a chemical kinetic model of autothermal pyrolysis to better understand the source of energy for the process.

Based on product differences between autothermal pyrolysis and inert pyrolysis (bio-oil, biochar, and gases), it appears the enthalpy for pyrolysis is being provided primarily by the oxidation of char, light oxygenates, and phenolic compounds. However, to confirm these products are in fact oxidizing at pyrolysis temperatures (400-600 °C) further experiments are needed. To accomplish this, model compounds representative of pyrolysis products such as acetic acid, levoglucosan, and phenolic monomers are partially oxidized at pyrolysis temperatures in a modified isothermal fluidized bed reactor. The condensed liquid product and non-condensable gases are analyzed via gas chromatography (GC) to determine products of partial oxidation and estimate kinetic data. Preliminary results show that these products begin to oxidize at temperatures below 400 °C, much lower than current published reaction mechanisms predict. Likewise, the role of incomplete combustion that does not generate CO_2 and H_2O appears to occur at these low temperatures and equivalence ratios. In particular, the lignin-derived phenolic components appear to incorporate oxygen into their structures during partial oxidation rather than generate non-condensable gases. Char oxidation studies were also performed using both thermogravimetric analysis (TGA) and fluidized bed experiments. Both methods indicate that the amount of alkali and alkaline earth metals (AAEMs) present in the biochar influence rates of reaction and product composition (CO vs CO_2). Determining the relationship between product oxidation at pyrolysis temperatures can identify not only the source of energy in biomass autothermal pyrolysis, but potentially expands autothermal processing to other applications as well.

576533 Structural and Chemical Changes in Plant Cell Walls during Early Stages of Thermal Deconstruction

Volatile products from lignocellulosic biomass thermal deconstruction processes—combustion, gasification, fast pyrolysis, and solvent liquefaction—have been well characterized, but the solid and liquid phase reactions that occur in the early stages of decomposition are largely unknown. Here we analyze the initial solid phase biomass thermal deconstruction reactions *in planta* and with high particle heating rates, creating the most accurate investigation to date of how these processes occur. Using a variety of instrumentation, we quantify the extent and relative rates of deconstruction, demonstrating that biopolymers resist the thermally energetic conditions to differing degrees, even when ensconced in biomass cell walls. Hemicellulose and the more frangible lignin components decompose and volatilize more readily than cellulose; this outcome temporarily enriches biomass with cellulose. These chemical changes additionally manifest in larger cell wall structure and mechanical property transformations. In all, our investigation concludes that these solid phase reactions strongly influence the production rates of volatile species and will require additional study before these processes can be modeled precisely or desired product yields potentially improved.

579943 Multiscale Simulation of Autothermal Pyrolysis of Biomass

The variability of biomass substantially exceeds that of conventional feedstocks such as coal due to inherent biocomplexities and inter-species diversity. The impacts of feedstock attributes which are unique to biomass, such as anisotropic, species-specific microstructure, highly variable ash and moisture content, and relatively large particle size distributions are well- recognized by the biomass conversion community. Problems associated with the complexity of biomass feedstocks have been acknowledged by the combustion and biopower communities. However, conventional simulation frameworks for thermochemical conversion of biomass feedstocks employ overly-simplified models that largely neglect the aforementioned complexities and/or apply generalized heuristics which limit their utility to the single feedstock and reactor system for which they were developed. In this talk will describe the development of a multiscale simulation approach that account for the complex size, shape, and chemistry of various biomass feedstock sources. The modeling framework is based on our previously-developed biomass particle pyrolysis simulations which account for the complex, anisotropic transport phenomena exhibited by biomass during thermochemical conversion processes. These particle scale models are coupled to reactor-scale simulations to account for bulk hydrodynamics and residence time distributions. In addition to these physics, we have implemented of a kinetic scheme suitable for biomass combustion and autothermal pyrolysis. I will also provide an update on our efforts to enable direct import of 3D image data of biomass particles into simulation environments and the implementation of adaptive mesh refinement to capture geometric changes that result from pyrolysis and combustion.

576446 Significant Increase in the Production of Sugars from Pyrolysis of Agricultural Residues By Application of Iron Sulfate Pretreatment

Fast pyrolysis of lignocellulosic biomass produces very little sugar or anhydrosugars compared to pyrolysis of pure polysaccharides because naturally occurring alkali and alkaline earth metals (AAEMs) in biomass catalyze the fragmentation of pyranose and furanose rings. Sugar yields can be dramatically increased by pretreating the biomass with sulfuric acid prior to pyrolysis, which passivates the catalytic activity of the metals by converting them into thermally stable salts. However, these AAEMs are also important to the depolymerization of lignin. Their passivation has the unintended consequence of slowing the rate of volatiles release from lignin, which results in a transient melt phase that agglomerates and fouls fluidized bed pyrolyzers. To overcome this problem, an alternative catalyst that is active toward lignin depolymerization but not pyranose ring fragmentation is needed. Pretreatment of herbaceous biomass with ferrous sulfate was found to have the desired effects of both passivating AAEMs and introducing iron that is catalytic toward lignin depolymerization. Conventional nitrogen-blown pyrolysis of pretreated corn stover increased the reactor weight hourly space velocity (WHSV), defined as biomass feed rate (kg/hr) per fluidized bed mass (kg), from 1 hr^{-1} with acid treatment to 4 hr^{-1} with iron sulfate. Substituting ferrous sulfate for sulfuric acid as pretreatment of corn stover had an even more dramatic increase in WHSV when operating the pyrolyzer under air-blown (autothermal) conditions, increasing from 3 $hr^{-1}$to 13 hr^{-1}. Sugar yields from the pyrolysis of corn stover increased from 0.5 wt% to 13 wt% on a biomass basis – a 26-fold increase through the use of the ferrous sulfate pretreatment. These advantages were combined to increase volumetric sugar productivity by two orders of magnitude, from 30 g/L/hr for conventional untreated corn stover to 3000 g/L/hr for autothermal ferrous sulfate treated corn stover.

579126 A Novel Fluidized Bed Model for Fast Pyrolysis Applications

Biomass fast pyrolysis has been identified as a viable method to produce bio-oil, a renewable alternative to petroleum-derived fuels and feedstocks. Fluidized bed reactors are ideal for pyrolysis due to their precise temperature control and high solid-solid heat transfer rates, and recent developments in autothermal operation have overcome the limitations inherent in supplying externally generated heat to the reactor itself. While many fluidized bed reactor models have been developed that accurately depict fluid-phase or heterogeneous reactions, they do not capture the intricacies of solid-phase reactions where heat transfer is the limiting factor.

This work presents a novel approach to the modeling of a fluidized bed reactor, focusing on the experience of the particles being pyrolyzed, which make up a minority of the bed mass. Due to the decoupling of solid- and vapor-phase reactions, this subset of particles can be modeled as a series of open control volumes surrounded by a uniform temperature heat source. The phenomena of heat transfer, chemical reaction, density loss, and elutriation are all captured in a format applicable within commercially available process simulators. The addition of partial oxidation reactions allows the reactor model to operate autothermally, with results supporting experimental findings on the performance enhancement derived from autothermal operation.

579942 CFD Investigation of Autothermal Biomass Pyrolizers

Autothermal biomass pyrolysis processes [1] are being developed by reaction engineers to address one of the main bottlenecks to the scale-up of biomass pyrolysis, which is the heat transfer limitation [1,2] due to the small heat transfer flux that can be sustained in the fluidized bed pyrolyzers used to conduct the process. Autothermal pyrolysis aims at addressing this limitation by allowing partial oxidation [1,3] of pyrolysis products, which allows the in-situ generation of heat needed to sustain the pyrolysis process.

The performance of autothermal pyrolizers may be affected by several design choices and by the composition of the biomass feed stream. The geometric configuration of the pyrolyzer and the location of the biomass feed may affect mixing of the biomass in the pyrolizer, impacting the residence time of the biomass and the product yield.

In order to investigate these effects, we have formulated a computational model to describe biomass fast pyrolysis. The model is based on the Euler-Euler multi-fluid model [4,5] with kinetic theory closures for polydisperse granular phases [6,7], supplemented by a comprehensive chemical kinetic mechanism [8–18] accounting for devolatilization, char combustion and gas-phase reactions. Results demonstrating the predictive capabilities of the computational models in comparison to experimental data of pyrolysis product yield will be presented.

The model was then used to study mixing of biomass and sand in a fluidized bed pyrolizer to understand the level of mixing obtained in the fluidized bed with different injection location

of the biomass. To this purpose, a mixing index insensitive to the spatial discretization of the computational domain was formulated, in order to systematically quantify the quality of the mixing of the granular mixture. Numerical experiments were first performed considering a cold flow in a laboratory scale fluidized bed with sand and biomass without accounting for the effect of eventual effects of the temperature distribution and of the chemical reactions.

The effects of heat transfer and chemical reaction are then introduced to investigate the impact of the location of the injection point of the biomass on the products in realistic operating conditions.

References

[1] J.P. Polin, C.A. Peterson, L.E. Whitmer, R.G. Smith, R.C. Brown, Process intensification of biomass fast pyrolysis through autothermal operation of a fluidized bed reactor, Appl. Energy. 249 (2019) 276–285. doi:10.1016/j.apenergy.2019.04.154.

[2] J. Proano-Aviles, J.K. Lindstrom, P.A. Johnston, R.C. Brown, Heat and Mass Transfer Effects in a Furnace-Based Micropyrolyzer, Energy Technol. 5 (2017) 189–195. doi:10.1002/ente.201600279.

[3] K.H. Kim, X. Bai, M. Rover, R.C. Brown, The effect of low-concentration oxygen in sweep gas during pyrolysis of red oak using a fluidized bed reactor, Fuel. 124 (2014) 49–56. doi:10.1016/j.fuel.2014.01.086.

[4] D.A. Drew, Averaged equations for two-phase flows, Stud. Appl. Math. L (1971) 205 – 231.

[5] D.A. Drew, Continuum Modeling of Two-Phase Flows, in: R. Meyer (Ed.), Theory Dispersed Multiph. Flow, Academic Press, 1983: pp. 173 – 190.

[6] J.T. Jenkins, S.B. Savage, A theory of the rapid flow of identical, smooth, nearly elastic spherical particles, J. Fluid Mech. (1983) 187 – 202.

[7] J.T. Jenkins, F. Mancini, Balance Laws and Constitutive Relations for Plane Flows of a Dense, Binary Mixture of Smooth, Nearly Elastic, Circular Disks, J. Appl. Mech. 54 (1987) 27–34. doi:10.1115/1.3172990.

[8] M. Calonaci, R. Grana, E. Barker Hemings, G. Bozzano, M. Dente, E. Ranzi, Comprehensive Kinetic Modeling Study of Bio-oil Formation from Fast Pyrolysis of Biomass, Energy Fuels. 24 (2010) 5727–5734. doi:10.1021/ef1008902.

[9] M. Corbetta, A. Frassoldati, H. Bennadji, K. Smith, M.J. Serapiglia, G. Gauthier, T. Melkior, E. Ranzi, E.M. Fisher, Pyrolysis of Centimeter-Scale Woody Biomass Particles: Kinetic Modeling and Experimental Validation, Energy Fuels. 28 (2014) 3884–3898. doi:10.1021/ef500525v.

[10] E. Ranzi, P.E.A. Debiagi, A. Frassoldati, Mathematical Modeling of Fast Biomass Pyrolysis and Bio-Oil Formation. Note I: Kinetic Mechanism of Biomass Pyrolysis, ACS Sustain. Chem. Eng. 5 (2017) 2867–2881. doi:10.1021/acssuschemeng.6b03096.

[11] R.S. Miller, J. Bellan, A generalized biomass pyrolysis model based on superimposed cellulose, hemicellulose and lignin kinetics, Combust. Sci. Technol. 126 (1997) 97–137. doi:10.1080/00102209708935670.

[12] C. Di Blasi, Heat, momentum and mass transport through a shrinking biomass particle exposed to thermal radiation, Chem. Eng. Sci. 51 (1996) 1121–1132. doi:10.1016/S0009-2509(96)80011-X.

[13] R. Radmanesh, Y. Courbariaux, J. Chaouki, C. Guy, A unified lumped approach in kinetic modeling of biomass pyrolysis, Fuel. 85 (2006) 1211–1220. doi:10.1016/j.fuel.2005.11.021.

[14] M.R. Rover, P.A. Johnston, L.E. Whitmer, R.G. Smith, R.C. Brown, The effect of pyrolysis temperature on recovery of bio-oil as distinctive stage fractions, J. Anal. Appl. Pyrolysis. 105 (2014) 262–268. doi:10.1016/j.jaap.2013.11.012.

[15] A. Williams, M. Pourkashanian, J.M. Jones, Combustion of pulverised coal and biomass, Prog. Energy Combust. Sci. 27 (2001) 587–610. doi:10.1016/S0360-1285(01)00004-1.

[16] R. Barranco, A. Rojas, J. Barraza, E. Lester, A new char combustion kinetic model 1. Formulation, Fuel. 88 (2009) 2335–2339. doi:10.1016/j.fuel.2009.02.005.

[17] C. Di Blasi, Combustion and gasification rates of lignocellulosic chars, Prog. Energy Combust. Sci. 35 (2009) 121–140. doi:10.1016/j.pecs.2008.08.001.

[18] G. Perkins, Underground coal gasification – Part II: Fundamental phenomena and modeling, Prog. Energy Combust. Sci. 67 (2018) 234–274. doi:10.1016/j.pecs.2018.03.002.

569566 Full-Scale Simulation, Energy and Environmental Analysis of an Actual LNG Chain

Full-scale Simulation, Energy and Environmental Analysis of an Actual LNG Chain

Mary A. Katebah[1], Mohamad M. Hussein[1], Abdur Rahman Shazed[1], Zineb Bouabidi[1], Abdulla Al-Hajri[2], and Easa I. Al-musleh[1]*

[1]Chemical Engineering Department, Qatar University

[2]Qatargas Operating Company

***Corresponding author.** Tel. +974 44034148. E-mail address: e.almusleh@qu.edu.qa (E. Al-musleh)

Natural gas (NG) is predominating the energy market as the main energy source worldwide, where it supplies over 20% of the total global energy demand.[1] The U.S. Energy Information Administration predicts that NG production will continue to grow from 83 to 93 billion cubic feet per day between 2018-2020.[2] This is credited to its lower carbon intensity and costs, compared to alternative fossil fuels. NG can be supplied to consumers by either pipeline transmission or as liquefied natural gas (LNG). While the former might be attractive over short distances, it offers limited capacity and flexibility compared to LNG.[3] Liquefaction allows for the storage and transportation of terawatt-hour amounts of NG at low pressures, making it the most cost-effective method of supplying NG to end-users over long distances. However, pre-treating and liquefying the NG involves complex energy-intensive steps that produce significant amounts of CO_2 and other harmful emissions (e.g. SOx, NOx, etc.). Additionally, LNG processes are associated with large quantities of water and waste discharges. Enhancing the efficiencies of the various units within the LNG chain is paramount in solving today's global dual challenge: "more energy with less carbon".[4] Furthermore, it results in substantial benefits including more LNG production, lower environmental impact, and higher profit. Process modeling and simulation is one of the quickest, predictive and cost-efficient tools to optimize a process. To date, no rigorous, comprehensive and validated process simulation model in the public literature exists for an entire actual LNG supply chain.

Japan is the world's largest importer, followed by China and South Korea.[5] In this work, we have used ProMax® and Aspen Plus® software to develop a rigorous base-case simulation model for an entire existing Qatar-Japan LNG supply chain. This included the upstream processing, liquefaction, utilities, transportation and regasification units. Two modes of operation were considered: without and with LNG carrier/receiving terminal loadings, referred to as holding and loading modes, respectively. A detailed methodology for simulating all the units within the LNG chain was developed while respecting process constraints, fuel balance and all product specifications. This approach can be extended towards simulating similar technologies within the process systems engineering field, generally, and the LNG value chain, specifically. The model was based on an actual process in Qatar, and its predictions match actual published data with less than 2% discrepancy. Simulation results were used to quantify the total plant fuel consumption and generation during the holding and loading modes. This was conducted by taking into account every single equipment within the LNG chain, including pumps, compressors, turbines, and steam generators, to name a few.

The model was also used to analyze the chain's performance through the first law of thermodynamics by calculating the energy efficiency of each unit, and their contribution to the chain's overall efficiency. The study also included a thorough environmental analysis to quantify CO_2, SO_x, NO_x emissions, in addition to process water and solid-waste discharge. Unlike the literature methodologies which utilize emission factors to quantify carbon dioxide emissions, rigorous simulation was used to obtain representative emission values. The study was concluded by assessing the relationship between the efficiency and environmental impact to highlight the areas that require further improvements. The model deemed to be a valuable tool that can serve as a basis for a plethora of reliable analyses and potential optimization studies.

References:

(1) Natural Gas. https://www.iea.org/topics/naturalgas/ (accessed Apr 1, 2019).

(2) Short-term Energy Outlook 2019. https://www.eia.gov/outlooks/steo/marketreview/natgas.php (accessed Apr 10, 2019).

(3) Patel, H.; Caswell, C.; Durr, C. North American LNG terminals: Options? *Hydrocarb. Process.* **2005,** No. July, 45–49.

(4) BP Energy Outlook 2019 edition. https://www.bp.com/content/dam/bp/business-sites/en/global/corporate/pdfs/energy-economics/energy-outlook/bp-energy-outlook-2019.pdf (accessed Apr 1, 2019).

(5) Shell LNG Outlook 2019. https://www.shell.com/promos/download-the-full-lng-2019/_jcr_content.stream/1550845695544/58c13c03d98ad9a29a6a27b5beab32d58b4ced34/lng-outlook-factsheet.pdf (accessed Apr 1, 2019).

(6) EIA: U.S. LNG export capacity to more than double by the end of 2019. https://www.worldoil.com/news/2019/1/1/eia-us-lng-export-capacity-to-more-than-double-by-the-end-of-2019 (accessed Apr 1, 2019).

560783 Replacing Steam with Methane in Thermal Cracking Processes

With a continued rise in its availability, shale gas represents a cheap raw material source for olefin production. Its major components include methane, ethane and propane, with methane making up over 60 mole% of all the components in the gas. Meanwhile, methane is mostly unreactive in thermal cracking reactions until temperatures of about 1000°C at ambient pressure. This particular feature of methane makes it particularly suitable to replace steam as a diluent in a thermal cracking process to produce olefins (such as ethylene) from paraffins (such as ethane and propane). Replacing steam with methane has a potential for cost savings in the construction and operation of a thermal cracking plant. Whereas liquid water has to be converted to steam before it can be used as a diluent in the steam cracking process and has high latent heat that must be supplied, methane need not undergo any such phase change before it can be similarly used in the present novel process.

Appropriate kinetics for methane, ethane and propane pyrolysis were selected from different sources in the literature to analyze the process for different operating temperatures ranging from 800 °C to 900 °C; and pressures of 2 Bar to 5 Bar using ANSYS Chemkin Pro. Two feed cases were used in the analyses. The first feed case has composition ratio of 0.795 mole methane, 0.147 mole ethane and 0.058 mole propane. The second feed case has composition ratio of 0.649 mole methane, 0.224 mole ethane and 0.127 mole propane. The compositions of these feed cases are typical of what are obtained from certain shale gas field regions in the United States.

Product distribution results show that methane is essentially an inert in the cracking process and basically serves to reduce the partial pressure of the ethane-propane components of the feed mixtures; thus, validating the novel idea of replacing steam with methane as a diluent. Also, ethylene is the major product obtained from the cracking process for all the cases and sets of operating conditions considered. This is similar to what would be obtained from a typical steam cracker. Required residence times for the process are also in the order of the ones required for the conventional steam cracker. Meanwhile, the process with methane as a diluent is simpler and has the potential for considerable savings in capital and operating costs of a thermal cracking plant compared to steam cracking.

574840 Comparative Study on Helium Recovery and Purification from LNG Liquefication Process

Helium is an indispensable chemical element, which is widely used in semiconductor applications, MRI scanning in medical industry, deep sea diving, weather forecasting, and welding due to its unique properties. However, the helium shortage is expected in near future because of increasing demand and the anxiety of supply. As a consequence, the helium production from alternative pathways has attracted the attention of industry. An increasingly available source of helium is from natural gas production, where helium could be collected from the end-flash unit from LNG liquefications process. It is considered as the most promising way of producing crude helium. Hitherto, there are three main types of helium extraction process have been proposed: flashing-based process, distillation-based process, and the integration of flashing and distillation process. Thus, the objective of this study is to perform a comparative evaluation based on rigorous simulations of these proposed helium extraction processes and give proper technical guidelines to achieve a successful helium extraction task. The simulation results indicate that the selection of the most suitable helium extraction process for an LNG plant will depend on the process, operation, and feedstock characteristics.

563133 Determination of the Failure Envelope of Kerogen Matrix By Molecular Dynamics Simulations

Shale gas is one of the most important unconventional resources. Hydraulic fracturing is the common method to create fractures in shale media for shale gas production. Generally, kerogen is the predominant component of organic matter in most shale oil/gas formations. Because kerogen has the potential to be the mechanically weak component in shale where the fracture initiates, it is critical to investigate its mechanical properties and failure mechanism. Failure envelope describes the locus of all shear and normal stresses at failure for a given rock material, which is often used as the criterion of rock failure for macroscopic simulations. Because kerogen is often dispersed in shale media in submillimeter range, it is impractical to accurately and systematically perform the mechanical tests on very small kerogen pieces to determine the failure envelope.

In this study, we provide a systematic study on the determination of the failure envelope of kerogen by molecular dynamics simulations. We construct various molecular structures of kerogen matrix, including type I, type II, and type III as well as the kerogen macromolecules with different maturities. The process of tension and compression at various states of stress are simulated. The stress-strain curves are obtained from the simulations, and Young's modulus, tensile strength, compressive strength, and fracture toughness are analyzed. The effect of pre-existing fracture and large pores are also investigated.

Our molecular simulations reveal that kerogen matrix experiences an elastic deformation first, and then a plastic deformation in both tension and compression. The shear failure approximately follows the Mohr–Coulomb criterion. Different kerogen types present significantly different failure modes. Type I kerogen shows ductile behavior, while type II and type III have a moderate brittle failure. However, all have less brittleness, weaker tensile strength, and weaker compressive strength than other minerals in shale. The fracture toughness is lower than the brittle minerals as well. The pre-existing fractures have a more significant effect on the mechanical properties than the large pores in the tensioning process. These effects will lead to different behaviors at the mesoscopic and macroscopic scale, which may result in different fracturing pressure.

In our work, we propose a practical method to determine the failure envelope of kerogen by molecular simulations for the first time. These results provide the essential parameters for mesoscopic and macroscopic simulations. The investigation at molecular scale is also critical to gain insight into the mechanisms of hydraulic and CO_2 fracturing for shale in the future work.

Modeling of gas hydrate particle decomposition for methane recovery under fluidized conditions

Airong Li [a, b, *], Dan Luo [a]

[a] *College of Chemistry and Chemical Engineering, Southwest Petroleum University, 610500, Chengdu, China*

[b] *Sichuan Collaborative Innovation Center of Marine Natural Gas Hydrate, Southwest Petroleum University, 610500, Chengdu, China*

Abstract: In this work, a fluidized bed was used for the first time for the process of methane recovery from natural gas hydrate. The decomposition of a spherical gas hydrate particle coupling with heat and mass transfer under fluidized conditions was quantitatively modeled. The results demonstrate that the decomposition is faster at a higher water flow velocity. The decomposition takes less time for the hydrate particle with a smaller initial size. Furthermore, the decomposition rates are improved by the convective heat and mass transfer between water and hydrate particle under fluidized conditions. Curve fitting and regression between methane recovery yield, water temperature, water flow velocity and initial particle size is also conducted, which renders quantitative insights in understanding the fundamental mechanism of gas hydrate decomposition under fluidized conditions.

Key words: natural gas hydrate, decomposition, fluidized bed, heat and mass transfer

1 Introduction

Conventional fossil fuels are being depleted due to overexploitation, and thus cannot satisfy the increasing energy demands of the society [1]. Therefore, more attentions have been paid to alternative energy sources such as natural gas hydrates, as the organic carbon in natural gas hydrates is twice as much as that in conventional fossil fuels [2, 3, 4].

Natural gas hydrates are widely distributed on the planet, especially in oceans [5]. Abundant methods have been proposed for methane recovery from hydrate decomposition, such as depressurization [6, 7, 8], thermal stimulation [6, 7, 8] and gas injection [9, 10]. The studies on hydrate decomposition are comprehensive. Kim et al. [11] firstly proposed the intrinsic kinetics of hydrate decomposition and demonstrated that the hydrate decomposition was affected by the temperature and the methane fugacity of gas phase. Based on Kim's research, Jamaluddin et al. [12] modeled the hydrate decomposition coupling with the intrinsic kinetics of hydrate decomposition with heat transfer. Selim et al. [13] modeled the heat and mass transfer for the process of hydrate decomposition in porous media and carried out an intensive study of hydrate decomposition. Many experiments on hydrate decomposition were also carried out [14-17]. These studies indicated that the heat and mass transfer was a key factor for hydrate decomposition.

The fluidized bed has been widely used in chemical and petrochemical industry due to its high efficiency of heat and mass transfer [18]. In a fluidized bed, solid particles are suspended in the fluid, which creates a fluidized condition. In this work, natural gas hydrate particles are suspended in the upward flowing heated water in the fluidized bed. Eventually, methane gas is decomposed and recovered from hydrate particles through fluidization. This process is shown in **Fig. 1.**

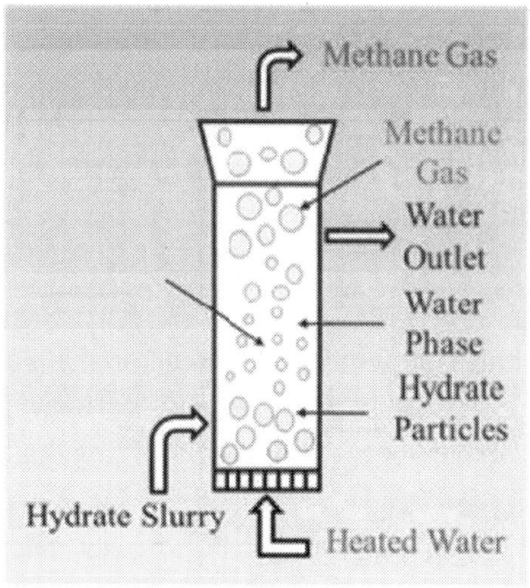

Fig. 1. Hydrate decomposition in a fluidized bed

Most of the existing literatures [11-17] on hydrate decomposition were under static conditions, and limited studies investigated the effect of flowing water on hydrate decomposition. Hamaguchi et al. [19] investigated the gas hydrate decomposition rate in flowing water and Xu et.al [20] studied the decomposition of natural gas hydrates in water lifting pipelines. However, these studies were under hydraulic conditions in which hydrate particles flow with water, rather than suspending in the flowing water. Research on heat and mass transfer between flowing water and hydrate particles was insufficient. Therefore, it is necessary to study the decomposition of hydrate particles with heat and mass transfer under fluidized conditions.

In this work, the decomposition of a single spherical gas hydrate particle under fluidized conditions was modeled, which was coupled with convective heat and mass transfer in the flowing water. The effects of water flow velocity, hydrate particle size and water temperature on heat and mass transfer were also investigated.

2 Model developments

The scheme of a single hydrate particle in the fluidized bed is shown in **Fig. 2**. In this proposed model, the decomposition film of hydrate particle was at the hydrate particle surface. The driving force of the decomposition is offered by the difference between the fugacity of methane at the phase equilibrium f_e and the fugacity of methane at the surface f_s. The heat and mass transfer of the hydrate particle decomposition under fluidized condition is considered. Q_r and M_r are the heat and mass flow in radial direction. T_i is the temperature and c_{gi} is the concentration of methane decomposed by hydrate particle in radial direction.

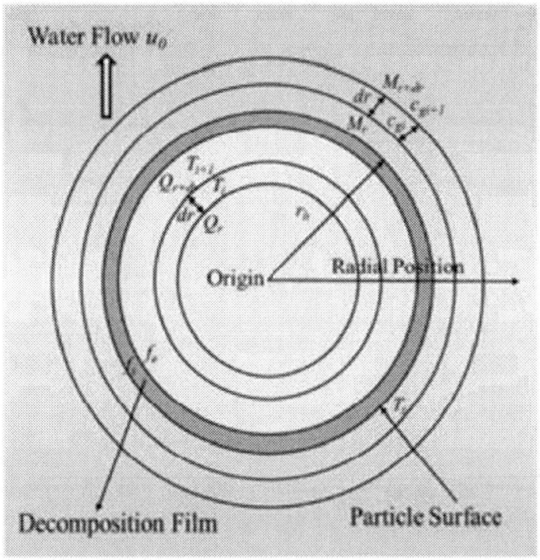

Fig. 2. Scheme of a spherical hydrate particle model

The basic assumptions for the model are as follows:

(1) The hydrate particle has a solid-state structure and would maintain its sphericity throughout the decomposition.

(2) The velocity u_0 and the temperature T_w of water flow are constants.

(3) Only methane is produced as the produced gas by hydrate decomposition and it does not affect the heat transfer.

(4) The heat and mass transfer occur only in radial direction of the spherical particle.

3. Model solution

3.1. Model Discretization

To solve the proposed model above, we divided the hydrate particle into infinitesimal gird points as shown in **Fig. 3**.

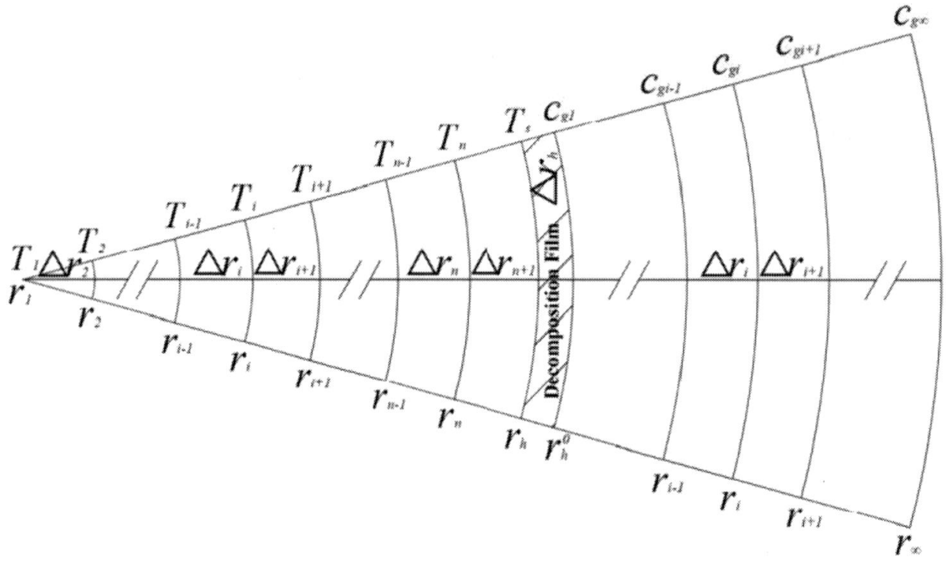

Fig. 3. Discretization of the gird points

The solving process of the model established above can be illustrated in **Fig. 4**.

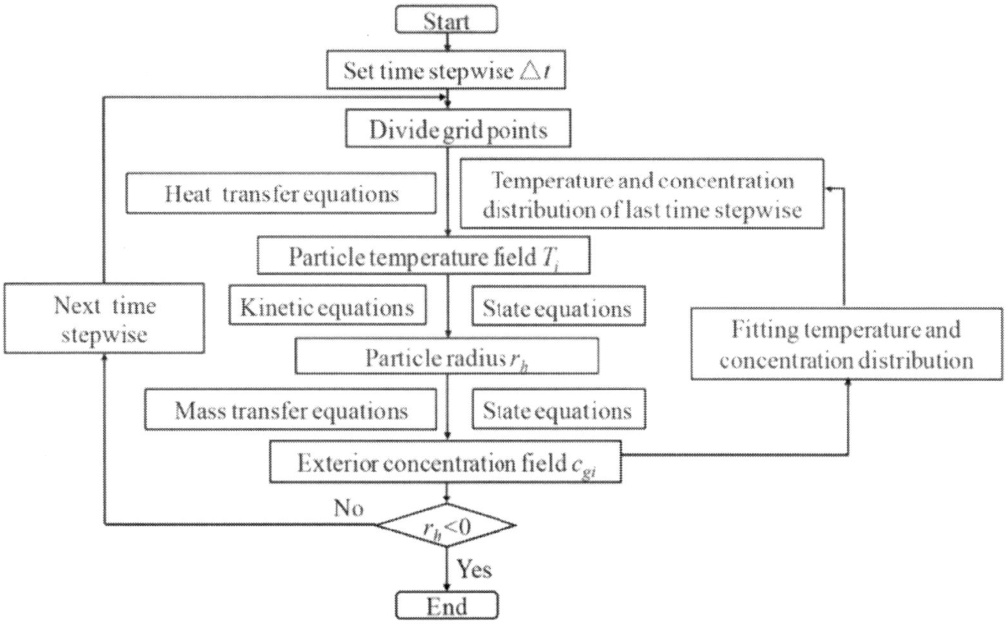

Fig. 4. The solving procedure of the model

4 Results and discussion

Water flow velocity is an important factor in the fluidized bed, and thus the effects of water flow velocity on heat and mass transfer are studied carefully. The effects of initial particle size and water temperature are also investigated.

Based on previous studies of hydrodynamics in the fluidized bed [21,22] and the particle size on hydrate decomposition [23], water flow velocity u_0 at 0.05 m/s, 0.10 m/s, 0.20 m/s and 0.40 m/s, and the initial particle radius r_0 of 0.05 cm, 0.10 cm, 0.20 cm and 0.40 cm are chosen for sensitivity analysis respectively. Considering the actual temperature distribution of seawater, the temperature of water flow T_w at 283.0 K, 288.0 K, 293.0 K and 298.0 K in the fluidized bed, are chosen for the model. The initial temperature of the hydrate particle T_h is assumed at 274.0 K.

The decomposition rate of the hydrate particle is determined by methane recovery yield (*MR*), which is defined by Eq. (1):

$$MR = \frac{Moles\ of\ methane\ decomposed\ from\ hydrate\ particle}{Total\ moles\ of\ methane\ in\ hydrate\ particle} \times 100\% \qquad (1)$$

4.1 Effect of water flow velocity

The effect of water flow velocity on methane recovery yield is shown in **Fig.6**. The methane recovery yield increases as water flow velocity increases. The methane recovery yields are about 51.92 %, 53.64 %, 54.78% and 55.54 % when the water velocities are at 0.05 m/s, 0.10 m/s, 0.20 m/s and 0.40 m/s in the first 60 s. This indicates that a higher water flow velocity would promote the hydrate decomposition, which might be explained by the reason that a higher water flow velocity enhances the convective heat and mass transfer. Therefore, the effect of water flow velocity on heat and mass transfer is further studied below.

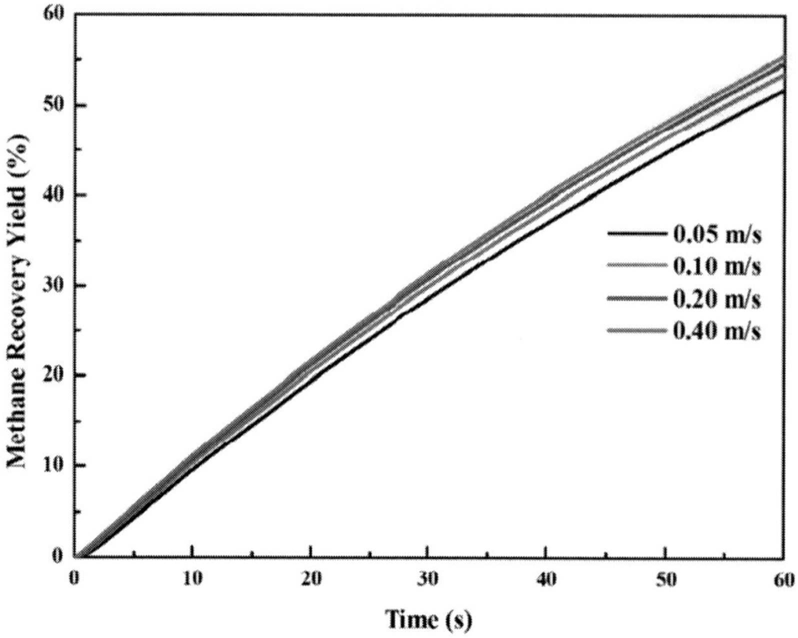

Fig.6 Effect of water flow velocity on methane recovery yield

4.2 Effect of initial particle size

The effect of the initial hydrate particle radius on methane recovery yield is shown in **Fig.7**. Methane recovery yields are about 84.43 %, 53.63 %, 29.32 % and 14.62 %, at the particle radius of 0.05 cm, 0.10 cm, 0.20 cm and 0.40 cm respectively during the first 60 s. Compared to the larger particle, the smaller particle has a higher methane recovery, because its larger specific surface area can promote the heat and mass transfer. Therefore, the smaller initial particle size can promote the hydrate decomposition. The effects of initial particle size on heat and mass transfer are further studied below.

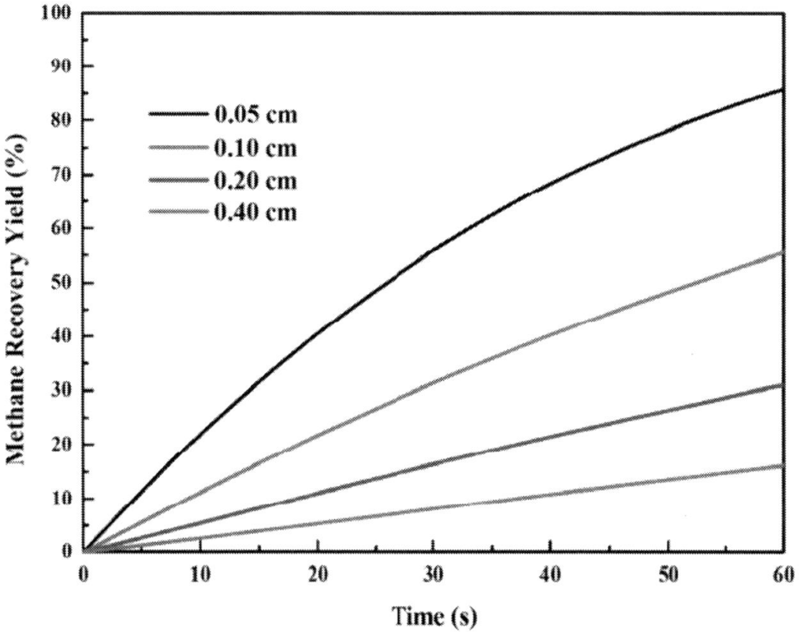

Fig.7 Effect of initial particle size on methane recovery yield

4.3 Effect of water temperature

The decomposition rate is directly related to the hydrate particle temperature, which is primarily determined by the water flow temperature. The effect of water temperature on methane recovery yield is shown in **Fig.8**. The methane recovery yield increases slowly with the decomposition time at lower temperatures of 283.0 K and 288.0 K, while it increases exponentially when the water temperature increases further. For the hydrate particle size of 2.0 mm at water temperature of 298.0 K and water flow velocity of 0.10 m/s, methane recovery yield can reach to 96.65 % in 60 seconds. Evidently, the effect of water temperature on hydrate decomposition is significant. Since higher temperatures can effectively enhance the heat flux in hydrate particles, the fugacity of methane at the phase equilibrium is higher, which renders the primary driving force for hydrate decomposition.

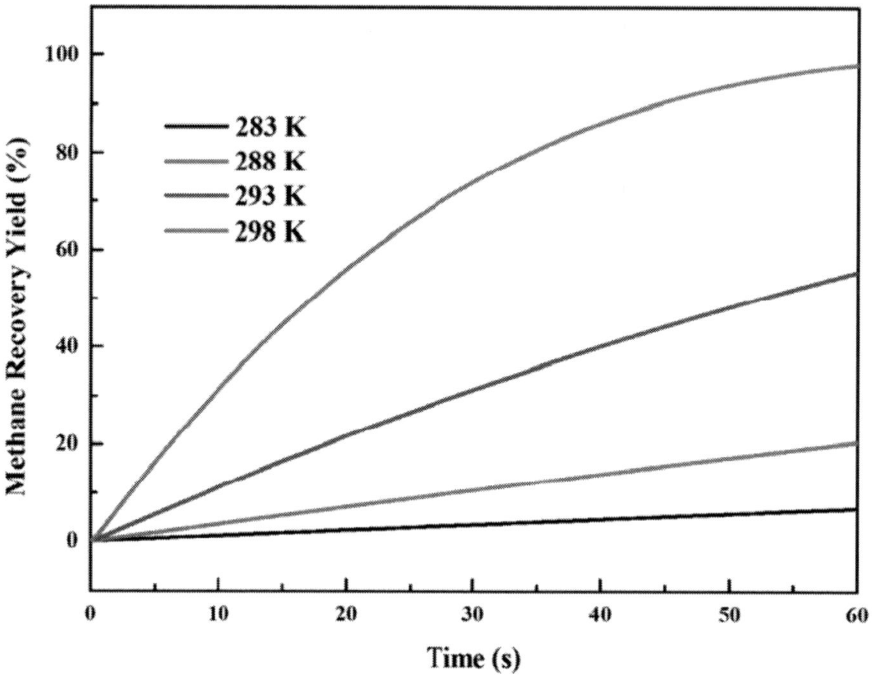

Fig.8 Effect of water temperature on methane recovery yield

4.4 Model regression

In order to quantitatively study the hydrate decomposition coupled with heat and mass transfer, curve fitting and regression is conducted. The correlation coefficients of methane recovery yield MR, time t (s), water temperature T_w (K), water flow velocity u_0 (m/s) and initial particle size d_0 (mm) are estimated from modeling data of hydrate decomposition under fluidized conditions. The data regression equation is presented in Eq. (2). The constants used in Eq. (2) are listed in **Table 1**.

$$MR = a_1 u_0^{b_1} d_0^{c_1} e^{\frac{d_1}{T_w}} t^2 + a_2 u_0^{b_2} d_0^{c_2} e^{\frac{d_2}{T_w}} t + a_3 u_0^{b_3} e^{\frac{d_3}{T_w}} \quad (2)$$

Table 1 The constants in the model of methane recovery yield

Parameters	Values
a_1	-1×10^{54}
b_1	0.1856
c_1	-2.1354
d_1	-39382
a_2	2.483×10^{26}
b_2	0.08455
c_2	-1.09206
d_2	-19085
a_3	-5.399×10^{25}
b_3	-0.5052
d_3	-19297

The fitting curve for methane recovery yield at water temperature of 290.0 K, water flow velocity of 0.15 m/s and initial particle size of 1.5 mm is shown in **Fig.9**. The fitting curve agrees well with the original modeling curve. Therefore, methane recovery yield under fluidized conditions might be estimated by Eq. (2).

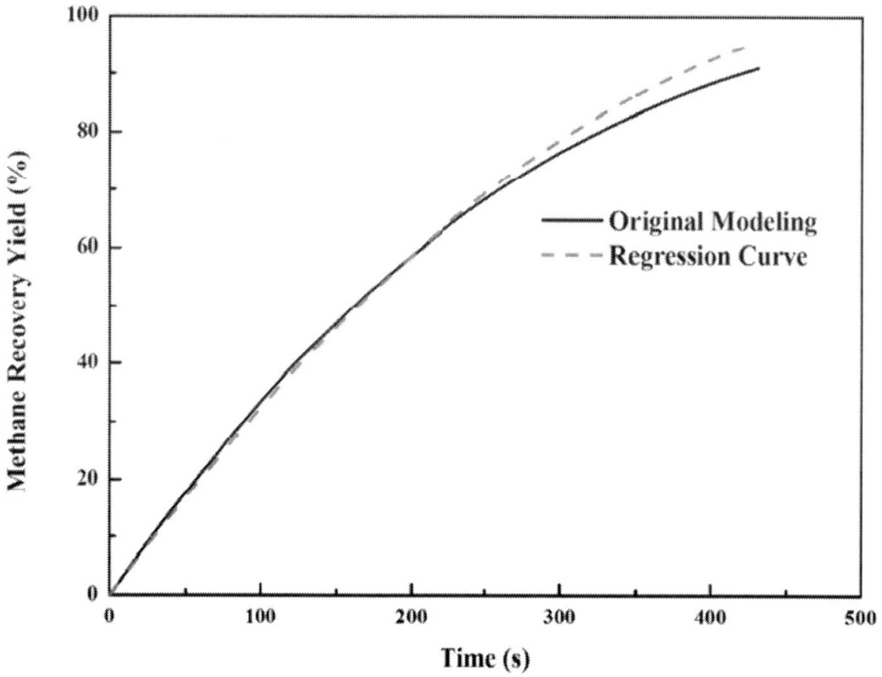

Fig.9 Fitting curve of methane recovery

5 Conclusions

In this study, a quantitative model is established to study the decomposition of a single natural gas hydrate particle with heat and mass transfer under fluidized conditions. The results showed that water temperature is the main factor affecting the dissociation of hydrate particles. For the hydrate particle size of 2.0 mm at water temperature of 298.0 K and water flow velocity of 0.10 m/s, methane recovery yield can reach to 96.65 % in 60 seconds. The increase in flow velocity and the decrease in initial particle size contribute to the enhancement of heat and mass transfer between hydrate particles and water. However, water flow velocity at small Reynolds number has no significant influence due to the heat transfer limitation of internal hydrate particle.

Acknowledgements

This work was supported by the Natural Science Foundation of China [grant numbers: 51704252] and Foundation of Educational Commission of Sichuan Province [grant numbers: 2017-263].

References

[1] Chong ZR, Yang SHB, Babu P, Linga P, Li XS. Review of natural gas hydrates as an energy resource: Prospects and challenges. Appl Energ 2015; 162: 1633-1652.

[2] Yin Z, Chong ZR, Tan HK, Linga P. Review of gas hydrate dissociation kinetic models for energy recovery. J Nat Gas Sci Eng 2016; 35: 1362-1387.

[3] Makogon YF, Holditch SA, Makogon TY. Natural gas-hydrates – A potential energy source for the 21st Century. J Petrol Sci Eng 2007; 56: 14-31.

[4] Makogon YF. Natural gas hydrates – A promising source of energy. J Nat Gas Sci Eng 2010; 2: 49-59.

[5] Demirbas A. Methane hydrates as potential energy resource: Part 2 – Methane recovery processes from gas hydrates. Energ Convers Manage 2010; 51: 1562-1571.

[6] Omidi M, Shahsavand A, Mohammadi HS. Modeling and simulation of hydrate Thermal Dissociation around Gas Production Pipe from Suboceanic Sediment. J Nat Gas Sci Eng 2016; 32:48-58.

[7] Jiang M, Fu C, Cui L, Shen Z, Zhu F. DEM simulations of methane hydrate exploitation by thermal recovery and depressurization methods. Comput Geothech 2016; 80: 410-426.

[8] Yang S, Lang X, Wang Y, Wen Y, Fan S. Numerical simulation of Class 3 hydrate

reservoirs exploitation using horizontal well by depressurization and thermal co-stimulation. Energ Convers Manage 2014; 77: 298-305.

[9] Shagapov VS, Khasanov MK, Musakaev NG, Duong NH. Theoretical research of the gas hydrate deposits development using the injection of carbon dioxide. Int J Heat Mass Tran 2017; 107: 347-357.

[10] Sridhara P, Anderson BJ, Garapati N, Seol Y, Myshakin EM. Novel technological approach to enhance methane recovery from Class 2 hydrate deposits by employing CO_2 injection. Energy Fuels 2018; 32: 2949-2961.

[11] Kim HC, Bishnoi PA, Heidemann RA, Rizvi SSH. Kinetics of methane hydrate decomposition. Chem Eng Sci 1987; 42: 1645-1653.

[12] Jamaluddin AKM, Kalogerakis N, Bishnoi PR. Modelling of decomposition of a synthetic core of methane gas hydrate by coupling intrinsic kinetics with heat transfer rates. Can J Chem Eng 1989; 67: 948-954.

[13] Selim MS, Sloan ED. Heat and mass transfer during the dissociation of hydrates in porous media. Aiche J 1989; 35: 1049-1052.

[14] Zhao J, Cheng C, Song Y, Liu W, Liu Y. Heat Transfer Analysis of Methane Hydrate Sediment Dissociation in a Closed Reactor by a Thermal Method. Energies 2012; 5: 1292-1308.

[15] Pang WX, Xu WY, Sun CY, Zhang CL, Chen GJ. Methane hydrate dissociation experiment in a middle-sized quiescent reactor using thermal method. Fuel 2009; 88: 497-503.

[16] Fitzgerald GC, Castaldi MJ, Zhou Y. Large scale reactor details and results for the

formation and decomposition of methane hydrates via thermal stimulation dissociation. J Petrol Sci Eng 2012; 94-95: 19-27.

[17] Li S, Zheng R, Xu X, Hou J. Energy efficiency analysis of hydrate dissociation by thermal stimulation. J Nat Gas Sci Eng 2016; 30: 148-155.

[18] Sang DK, Yong K. Heat and mass transfer in three-phase fluidized-bed reactors—an overview. Chem Eng Sci 1997; 52: 3639-3660.

[19] Hamaguchi R, Nishimura Y, Inoue G. Gas Hydrate Decomposition Rate in Flowing Water. J Energy Resour Technol 2007; 129: 102-106.

[20] Xu H, Kong W, Yang F. Decomposition characteristics of natural gas hydrates in hydraulic lifting pipelines. Nat Gas Ind 2018; 38: 129~136. (in Chinese)

[21] Rampure MR, Buwa VV, Ranade VV. Modelling of Gas-Liquid/Gas-Liquid-Solid Flows in Bubble Columns: Experiments and CFD Simulations. Canadian J Chem Eng 2003; 81: 692-706.

[22] Li W, Zhong W. CFD simulation of hydrodynamics of gas–liquid–solid three-phase bubble column. Powder Technol 2015; 286: 766-788.

[23] Misyura S Y. Effect of diameter of granules on dissociation of methane hydrate. J Eng Thermophys 2018; 27(2):191-195.

575460 CFD Based Dynamic Boil-Off Gas Study during LNG Laden Voyage

Liquified Natural Gas (LNG) is an increasingly important clean energy source, a widely used alternative fuel and is transported all over the world. Currently, the boil-off gas (BOG) generation, minimization, and utilization during LNG storage and shipping are still major concerns for the industry. To address these concerns, the first and also the most important step is to accurately quantity BOG generation amount with consideration of all aspects of root causes. Generally, during the long-distance marine transportation of LNG, two important factors for BOG generations have to be accounted: heat leakage and vibration impact. Many studies have been reported on heat leakage impact on BOG generation. But quite few addresses the impact of LNG tanker vibrates induced by sea waves during the LNG laden voyage. Our previous study on LNG road tanker has indicated that LNG moving caused by mechanical vibrations can also contribute to some extent for the BOG generation. Compared with road transportation, marine transportation will experience stronger vibration and sloshing of fluid inside storage tanks. Thus, the vibration effect on BOG generation during LNG marine transportation cannot be neglected.

In this study, the dynamic BOG generation caused by both vibration and heat leakage during its marine transportation has been investigated quantitively with the power of both computational fluid dynamics (CFD) and process dynamic simulation, the commercial CFD code ANSYS Fluent was V18.1 was applied to develop a multiphase mixture model and simulated fluid displacement caused by vibration and sloshing during laden voyage, where a user-defined function (UDF) was compiled to model the fluid moving inside the tank, energy generated by vibration and sloshing was exported and coupled with heat leakage modeling via process dynamic simulation to quantify the amount of the total dynamic BOG generation. This study could lay a solid foundation for further study on BOG generation minimization or utilization during LNG marine transportations.

561620 Static Boil-Off Rate (BOR) Implication on Boil-Off Gas (BOG) in an LNG Regasification Terminal

Liquefied natural gas (LNG) is stored in the tank at its bubble point and any heat leak boils the natural gas (NG) to form boil off gas (BOG). BOG generation undone the compression work required for NG liquefaction. Thus, keeping the BOG to minimum increases the LNG value chain economics. Though BOG depends on the amount of heat leak, assessing it from the LNG storage tank is often based on the static value of BOR (Boil off rate), suggested by the tank vendors. Static BOR value depends on the worst case heat leak in the tank and always over predict the BOG generation because it assigns all the heat leak into the tank for BOG generation disregarding the heat leak to vapor that causes vapor expansion. This study investigates the impact of static BOR on BOG over prediction at different liquid levels in the tank. The analysis revealed that BOR varies from 0.012wt% per day at 80% tank fill to 0.12wt% per day at 10% tank fill. This means that for a 90m high tank, every 1meter decrease in LNG level over predicted BOR by 0.004%. Thus, instead of using static value of BOR for BOR predictions a variable value of level dependent BOR give more accurate predictions for BOG estimation and must be adopted.

553853 Prediction of the Fugacity of Hydrogen Gas Dissolved in Organic Solvents with Emphasis on the High-Temperature High-Pressure Subcooled Liquid State

The so-called Nelson Chart is used to determine susceptibility of steel piping to High Temperature Hydrogen Attack (HTHA) based on the type of steel, temperature and the amount of hydrogen in a process stream. Although the chart is presented in terms of the partial pressure of hydrogen in the stream, this variable is unsuitable for evaluating streams where you have subcooled liquid and no vapor phase is present. In this situation the desired thermodynamic property reflecting the concentration of hydrogen in the stream is the fugacity of dissolved hydrogen. This study investigated the accuracy of rigorous methods of calculating the fugacity of hydrogen dissolved in hydrocarbon streams typical of the petroleum refining industry. Several equations of state were used to calculate the fugacity of hydrogen dissolved in saturated and subcooled liquid hydrocarbon phases. The predictions of two correlative equations of state (EOS), Peng-Robinson and PC-SAFT, were compared to experimental data for the fugacity of dissolved hydrogen. In addition, values calculated using two other 'predictive' methods, PSRK and VTPR, were examined.

The dissolved hydrogen fugacity data were obtained by integrating binary experimental data for the partial molar volume (PMV) of hydrogen dissolved in organic non-polar solvents. A literature survey identified hydrogen high-temperature high-pressure PMV data for six solvents: n-pentane, 2,3-dimethylbutane, benzene, cyclohexane, n-octane, 1,4-diethylbenzene, m-xylene and n-decane. As a preliminary step, the Peng-Robinson and PC-SAFT EOS were fit to binary vapor-liquid equilibrium (VLE) data for hydrogen with the respective solvents. The resulting fits of these two equations, as well as the predictions of the three other models were compared to the VLE data. The results of the comparison indicate that the correlative models adequately represent the VLE data, albeit with large values of the binary interaction coefficients. The 2 fitted models and predictive models were then used to calculate values of the partial molar volume and fugacity of hydrogen in the binary systems over a wide range of temperature, pressure and hydrogen concentration.

Using the EOS fit to VLE data and the two predictive models, additional calculations were performed to predict values of hydrogen fugacity in multicomponent liquid phase streams typical of the refining industry with products such as naphtha, kerosene and diesel. The results of this study are useful in validating short-cut methods for predicting hydrogen fugacity in subcooled streams to be used in HTHA susceptibility.

572726 Pseudo-Component Techniques for Prediction of Thermophysical Properties of Hydrocarbon Mixtures and Fuels up to High Temperatures and Pressures

Experimental property measurements for fuels (*e.g.*, gasoline, kerosene, jet, and diesel) at high temperature and high pressure (HTHP) conditions are expensive and time consuming. These limitations can be overcome using an equation of state (EoS) to predict mixture properties beyond the range of experimental observations. Although fuels are complex mixtures of hydrocarbon compounds with properties and composition that vary due to quality specifications, only a few composition-based approaches have been proposed to model their properties. However, these techniques have been limited by the need for experimental measurements to fit the EoS parameters or requiring complex compositional characterization to define multiple pseudo-components.

Here, we present pseudo-component techniques based upon the perturbed-chain statistical associating fluid theory (PC-SAFT) EoS of Gross and Sadowski (*IECR*, **40**, 1244, 2001) to predict thermodynamic and transport properties of hydrocarbon mixtures and fuels up to HTHP conditions. We use group contribution approaches to correlate and calculate the model parameters, treating the mixtures as a single pseudo-component and eliminating the need for binary interaction parameters. Only two to three measured or calculated mixture properties are required as inputs: the number averaged molecular weight, hydrogen to carbon ratio of the mixture, and optionally for transport properties, a single data point at a chosen reference point to fit one of the model parameters. Transport properties (*i.e.*, viscosity and thermal conductivity) are predicted by combining PC-SAFT with the entropy scaling approach of Rosenfeld (*Phys. Rev. A*, **15**, 2545, 1977). Thermophysical properties are predicted for hydrocarbon mixtures and jet and diesel fuels up to 600 K and 3,500 bar with average mean absolute percent deviations of 1, 9, and 2 % for density, viscosity, and thermal conductivity, respectively, compared to experimental data.

569958 Development of a Thermodynamic Model and Multiphase Equilibrium Flash for Prediction of Crystallization in LNG Processes

Development of a thermodynamic model and multiphase equilibrium flash for prediction of crystallization in LNG processes

Nevin Gerek Ince[1],*, Freddy Garcia[2], Jean-Jacques Bartuel[3], Seiya Hirohama[1], David Bluck[1]

[1]*AVEVA Software*

26561 Rancho Parkway South, 92630 Lake Forest, California, United States

[2]*TOTAL*

2, place Jean Millier. 92078 Paris La Défense CEDEX France

[3]*Technip*

6-8 Allée de l'Arche

92973 Paris La Défense Cedex France

**Corresponding/Presenting author: nevin.gerekince@aveva.com*

Abstract

Worldwide production of natural gas is growing, mainly through the development of new shale gas fields, using hydraulic fracturing. The distance between production and consumption sites is increasing, making pipeline transmission of natural gas less economically attractive than liquid phase transportation as Liquefied Natural Gas, or LNG. In the process of liquefying natural gas, the gas feed has to be cooled to very low temperatures of around 100K or -173°C to keep the vapor pressure low enough for safe containment. Hydrocarbons heavier than the main constituents of natural gas, methane, ethane and propane, even in trace amounts, can result in crystallization in the liquefaction process. This can lead to plugging and potential damage to expensive equipment, such as cold-box heat exchangers. Operators of LNG plants want to be able to push operating conditions as close to the limits as possible, without risking the formation of solids in the process as feed stream compositions change.

Simulating LNG productions requires an accurate and robust thermodynamic framework which can handle solid phase at the cryogenic conditions. For this goal, we gathered the latest experimental data, and critically examined them. Thermodynamic models were systematically investigated applying Cubic Equations of State (Soave-Redlich-Kwong in this study), tuning pure component data and binary interaction parameters and alpha formula considering thermodynamic constraints. We updated correlations for solid fugacity, and density models. A robust multi-phase – V(mL)(nS)E - flash algorithm was established for vapor-liquid-liquid-solid (VLLSE) equilibrium by applying phase stability analysis. The revamped thermodynamic methods and data allowed us to accurately predict the multiple phase boundaries that can be present when cooling a typical natural gas mixture. Simulation results were also compared against experimentally measured vapor-liquid-solid equilibrium data.

559335 Advances in Modeling Asphaltene Precipitation and Deposition

Asphaltene precipitation and subsequent deposition is a flow assurance problem that threatens the continuous production of crude oil. Asphaltenes constitute the heaviest and the most polar fraction of the crude oil, which can precipitate in the wellbore due to changes in temperature, pressure and composition. The formation of asphaltene deposits can lead to significant downtime, deferred production and expensive remediation procedures.

In this talk we will present the state-of-the-art modeling methods to forecast the occurrence and the magnitude of asphaltene precipitation and deposition. We will provide some field examples and important best practices related to this important problem, faced by oil companies worldwide.

569731 A Study of Unexpected Autoignition Temperature Trends for Pure *N*-Alkanes

An experimental study of minimum autoignition temperatures (AIT) reveals unexpected behavior of pure, high-carbon-number *n*-alkanes (C16 and larger) not predicted by current AIT prediction methods. The well-documented behavior of *n*-alkanes shows AIT starting higher for the smaller members of the family and then decreasing to a roughly constant value with little change between members of the family from C7 and larger. Our measurements show that values start to increase again at C16 and show a dramatic increase between C25 and C26.

Measured values for several *n*-alkanes using the ASTM E-659 method are reported and compared to data from other sources. The impact of measurement method on AIT values is discussed. A mechanistic explanation of the unexpected behavior is proposed. Finally, the implications of this behavior on estimation techniques and on the trend with carbon number for other chemical families are discussed.

565697 Investigation of the Flammability Limits of Petroleum Liquids

Investigation of the Flammability Limits of Petroleum Liquids

Controlling vapor concentrations of flammable liquids is a major consideration in safety analysis. This presentation discusses a study of the explosive behavior of a kerosene sample used for a laboratory desulfurization study. The first step is reliable and accurate estimation of the flammability limits. The kerosene is assumed to be composed of saturated hydrocarbons so that the accurate flammability limits reported in the DIPPR database [1] could be used for the pure components. The estimation of the mixture flammability limits uses the model proposed by Liaw [2], and thoroughly tested by Vidal, Rogers and Mannan [3]. The model has been tested for relevant hydrocarbon mixtures by using data reported in the literature [4, 5]. The validated model predicts that the vapor space in the laboratory reactor is indeed in the flammable region, and hence the head space of the reactor should be open and a purge system is necessary to keep the concentrations of the flammable species at a safe level (about a factor of 5 below the lower flammability limit).

References

1. http://www.aiche.org/dippr
2. Liaw, H.-J.; Chiu, Y.-Y. The Prediction of the Flash Point for Binary Aqueous-Organic Solutions. *J Hazard Mater.*, **2003**, A101, 83–106.
3. Vidal, M.; Rogers, W. J.; Mannan, M. S., Prediction of Minimum Flash Point Behaviour for Binary Mixtures. *Process Safety and Environ. Prot.*, **2006**, 84, 1-9.
4. Lenoir, J. M. Predict Flash Points Accurately. *Hydro. Proc.*, **1975**, 54, 95-99.
5. Zabetakis, M. G., Flammability Characteristics of Combustible Gases and Vapors, Bureau of Mines, Washington, D.C. **1965**.

555610 A Comparative Review of Thermophysical Properties and Characteristics of Biofuels and Petrofuels

With limited sources of conventional petroleum reserves and due to environmental impacts of petroleum derived fuels (petrofuels), production and uses of biofuels (fuels produced from biomass) are on the rise. It is projected that by 2050, biofuels contribute to about 30% of world transportation fuels. The main purpose of this presentation is to demonstrate and to compare the nature and characteristics as well as thermophysical properties of biofuels with petrofuels.

Although there are many types of biofuels in the forms of gaseous (such as biomethane or biohydrogen) as well as in liquid and solid forms but the two types of biofuels that have been commercialized are biodiesel and bioethanol and both are in liquid forms. In addition to regular biodiesel produced from alcohols, biodiesel can be produced from hydrotreated vegetable oil (HVO) which hydrogen is used as a catalyst to convert vegetable oil to biodiesel [1]. Properties of these biofuels (produced from different paths and different raw materials) are compared with specifications of different petrofuels such as gasoline, diesel, jet fuel, and light gas-oil used as transportation fuels. Physical properties (ie, density , boiling range and transport properties), thermal properties (ie, heating value, heat capacity), properties related to engine performance (cetane number, fuel consumption), safety (ie, flash point), cold conditions (pour point), fuel quality (ie, H/C ratio, aromatic, sulfur and oxygen contents) as well as fuel impacts on the environment (ie, CO, NO and sulfur emissions) are discussed in this presentation.

[1] Riazi, M. R. and Chiaramonti, D. (editors), Biofuels Production and Processing Technology, CRC Press, 690 pages, October 2017.

568903 Prediction of Multicomponent Gas Adsorption Equilibria and Heat of Adsorption from Pure Component Isotherms with Adsorption NRTL Model

Abstract

In gas adsorption equilibria, successful engineering thermodynamic models should meet five criteria: (i) thermodynamic consistency, (ii) least numbers of physically significant adjustable parameters, (iii) applicability in both pure and multicomponent adsorption, (iv) predictivity of multicomponent adsorption equilibria from its pure component isotherms, and (v) predictivity of isosteric heat of adsorption for pure and multicomponent systems [1, 2]. Numerous researchers have proposed empirical and semi-empirical models such as Langmuir, Toth, and multisite Langmuir to represent pure and multicomponent gas adsorption equilibria [3-6]. However, to the best of our knowledge, none of the models fulfill all these requirements. To resolve this issue, recently we developed an adsorption non-random two liquid theory (aNRTL) activity coefficient model to account for specific adsorbate-adsorbent interactions in the adsorbate phase. In this work, we show the aNRTL model successfully represent pure component isotherms as well as isosteric heats of adsorption. We further show the estimated model parameters from pure component isotherms successfully predict the mixed-gas adsorption equilibria and the associated isosteric heats of adsorption. Meeting the five criteria for a successful engineering thermodynamic model for gas adsorption equilibria, the aNRTL model should serve well as a unifying thermodynamic framework to correlate and predict adsorption isotherms and to support modeling and simulation of adsorptive separation processes.

References

[1] S. Sircar, "Role of adsorbent heterogeneity on mixed gas adsorption," *Industrial & Engineering Chemistry Research,* vol. 30, pp. 1032-1039, 1991.

[2] P. M. Mathias, R. Kumar, J. D. Moyer, J. M. Schork, S. R. Srinivasan, S. R. Auvil, *et al.*, "Correlation of multicomponent gas adsorption by the dual-site Langmuir model. Application to nitrogen/oxygen adsorption on 5A-zeolite," *Industrial & Engineering Chemistry Research,* vol. 35, pp. 2477-2483, 1996.

[3] S. J. Bhadra, A. D. Ebner, and J. A. Ritter, "On the use of the dual process Langmuir model for predicting unary and binary isosteric heats of adsorption," *Langmuir,* vol. 28, pp. 6935-6941, 2012.

[4] D. D. Do, *Adsorption Analysis: Equilibria and Kinetics* vol. 2: Imperial College Press, London, 1998.

[5] A. Myers and J. M. Prausnitz, "Thermodynamics of mixed-gas adsorption," *AIChE Journal,* vol. 11, pp. 121-127, 1965.

[6] O. Talu and I. Zwiebel, "Multicomponent adsorption equilibria of nonideal mixtures," *AIChE Journal,* vol. 32, pp. 1263-1276, 1986.

573445 Development of Robust Models for the Prediction of Reid Vapor Pressure (RVP) in Fuel Blends and Their Application to Oxygenated Fuels

The growing demand for inexpensive, more environmentally friendly fuels has sparked research into fuel properties that affect fuel production, handling and engine efficiency. Reid Vapor Pressure (RVP) has been shown to be one such fuel property. Having models that can accurately predict the RVP of a fuel quickly and robustly can help in the development of novel fuels, increase engine efficiencies and value for refiners in their existing operations. Due to non-linear blending effects, especially when oxygenates are involved, current methods often fail to predict RVP correctly. New methods are required which can accurately take non-linear blending effects into account. To that end, models were developed that can predict the RVP of complex Blendstocks for Oxygenated Blending (BOB) using the SAFT-γ-mie Equation of State (EOS). This approach captures the underlying physics and has been shown to be reliable for the prediction of thermophysical properties of complex fluids. The models increase in fidelity and accuracy, with increasing complexity, i.e. number of compounds included, to construct. Three molecular based "surrogate" models are presented. Each model uses a set of simple rules to determine which components must be included, with the small surrogate model including the fewest number of components, and the large surrogate model including the largest number of components. In order to assess the accuracy of each model, seven BOBs with a wide range of known RVPs and representing a wide range of chemical compositions were evaluated. The present absolute average deviation %AAD, averaged over these seven BOBs, is calculated to be 15.09%, 9.49%, and 3.66%, for the small, moderate, and large surrogate models, respectively. To test the performance of oxygenates added to the system, two of the BOBs, a winter and summer, were blended with several oxygenates of varying molecular class. The %AAD for the summer BOB data is 19.72%, 14.90%, and 3.48% for the small, moderate, and large surrogate model, respectively. The %AAD for the winter BOB data is 15.50%, 12.16%, and 2.58% for the small, moderate, and large surrogate model, respectively. In addition, excess thermodynamic properties for several binary systems were examined. These were used to create a physical model of RVP, which can be used to inform future fuel research. Lastly, using the large surrogate model, the RVP was predicted for a suite of oxygenates for blends of 10, 20, and 30%, by weight. Although the majority of oxygenates tested decreased the RVP, the branched alcohols, and the ketones showed the largest decreases, suggesting possible refinery applications through the decrease of fugitive losses.

557160 The Effect of Bio-Derived Additives on Fatty Acid Methyl Esters for Improved Biodiesel Cold Flow Properties

Biodiesel is a promising replacement for petroleum based diesel in a number of different applications. Although biodiesel offers numerous advantages over petrodiesel, one of its major disadvantages is its inferior cold flow properties. Additives are often added to biodiesel to remediate issues with cold flow and other problematic properties. This work explored the feasibility of using compounds derived from bio-based sources as an additive for biodiesel. Specifically, the additive's efficacy in reducing the cloud point of model biodiesels was analyzed using differential scanning calorimetry. Each of the additives studied depressed the biodiesel cloud point, although most of the additives showed diminishing returns above a certain weight percent. For many of the additives, the mechanism of cloud point depression was found to be dilution. For these additives, their efficacy was hampered by self-association, a phenomenon confirmed by thermodynamic modeling. However, tripropionin and dimethyl azelate, molecules of similar size to the FAMEs in the biodiesel studied in this work, interfered with the development of a critical nucleus, causing more significant depressions in the cloud point. Overall, a thermodynamic model based on ideal solution theory was able to accurately predict cloud points for most mixtures studied. Large deviations were observed for systems with significant amounts of self-association and/or cocrystallization.

559408 Crude Oil Compositional Analysis Methods and Their Impact on Thermophysical Properties

Crude oil composition can be useful when extracting, transporting, or refining crude oils since the chemical makeup affects properties linked to the safety, transportation requirements, and value of the material. Determining composition requires a combination of several analysis methods due to the wide range of constituents, from dissolved gases (N_2, CH_4, CO_2) to light- and mid-range hydrocarbons to heavy ends (C_{30}+). These analytical methods, each optimized to a fraction or boiling range cut of the original sample, must be followed by a numerical or physical recombination step to recreate a "whole oil" composition in which each constituent is quantified on the basis of its mass, volume, or mole percentage of the whole sample. In some cases, the limitations of the individual analytical methods propagate to the recombined whole oil. These limitations become clearer as Equation of State (EOS) model calculations are compared to direct property measurements, eliciting the need to carefully select analysis methods. The analysis presented here was performed on tight crude oils sampled mid-stream in North Dakota and Texas as part of the Crude Oil Characterization Research Study.

The Crude Oil Characterization Research Study (funded by US DOE, DOT, and Transport Canada: http://energy.gov/fe/articles/crude-oil-characteristics-research, https://www.theatlantic.com/technology/archive/2018/02/the-great-crude-oil-fireball-test/552029/), seeks to understand the crude oil property differences between conventionally produced and tight oils. Recent activities analyzed the effects of sampling containers/techniques (ASTM D3700, GPA 2174, ASTM D8009, ASTM 4057) and sampling locations (Texas vs. North Dakota) on crude oil thermophysical property measurements. Different compositional measurement techniques (GPA 2103, GPA 2177, ASTM D7900, ASTM D7169, ASTM D8003, GOR flash gas) have been used to calculate these properties as well, providing insight into the features and limitations of the methodologies.

Sandia National Laboratories is a multimission laboratory managed and operated by National Technology & Engineering Solutions of Sandia, LLC, a wholly owned subsidiary of Honeywell International Inc., for the U.S. Department of Energy's National Nuclear Security Administration under contract DE-NA0003525.

569963 Quaternary Phase Equilibrium Data of Quaternary Water + (Ethanol and 2-Propanol) + Isooctane System

Quaternary phase equilibrium data of quaternary Water + (Ethanol and 2-Propanol) + Isooctane system

Danielle L. de Klerk, Cara E. Schwarz[1]*

[1]*Department of Process Engineering, Stellenbosch University, Banghoekweg, Stellenbosch, 7600, South Africa*

**email: cschwarz@sun.ac.za, Tel: +27 21 8084485*

Introduction

Worldwide there is a need for anhydrous low molecular mass alcohols, such as ethanol and 2-propanol. These alcohols have many industrial applications, such as in biofuel, high-purity cleaning solvents or as fuel additives to enhance the octane rating of fuel. Very often the alcohols are produced as dilute aqueous mixtures in large scale operations, such as fermentation or by-products of the Fischer-Tropsch process. However, ethanol and isopropanol both form binary azeotropes with water and thus normal distillation is unable to produce anhydrous alcohols and advanced separation techniques are required.

Heterogeneous azeotropic distillation is a well-established technique capable of separating alcohol and water mixtures. The addition of a third component, the entrainer, overcomes the azeotrope by forming a ternary heterogeneous azeotrope with the water-alcohol system. Industrially, benzene has been used as a very successful entrainer in the separation of water and alcohols.

Unfortunately, benzene is carcinogenic and toxic. Thus, a strong incentive and drive exists to replace benzene as an entrainer. Many alternatives have been proposed, including isooctane. Several studies have been done to show that isooctane is a possible suitable alternative as it forms a ternary heterogeneous azeotrope with the water and ethanol or 2-propanol (Font, et al., 2003; 2004). Furthermore, if the alcohols are to be used as fuel additives, traces of isooctane are not detrimental to the process.

Existing thermodynamic models struggle to accurately predict the phase behaviour of such systems, and as such reliable phase equilibria data are required to understand the system's phase behaviour, for use in modelling correlation and subsequently in process design. Phase equilibria data are available for the ternary water + alcohol + entrainer systems, but very little, if any, research has been done into the case where multiple alcohols are present. Investigation into such systems is warranted as there is often more than one alcohol present in the dilute aqueous stream undergoing purification, such as from Fischer Tropsch reaction water, and the presence of the second alcohol may influence the existing ternary phase behaviour.

The aim of this contribution is therefor to investigate the phase equilibria of the quaternary water + ethanol + 2-propanol + isooctane system by presenting (1) liquid-liquid equilibrium (LLE) data for a range of temperatures; (2) vapour-liquid equilibrium (VLE); and (3) vapour-liquid-liquid equilibrium (VLLE) data of the system at three 2-propanol: ethanol ratios at 101.3 kPa.

Methodology

LLE experimental methodology

The LLE phase equilibrium data measurements were conducted in shakeflasks. A known composition was loaded and allowed to reach thermal equilibrium, while shaken regularly. The mixture was left for a minimum of fifteen hours to separate, before samples were drawn from each phase. The ratio of the organic components in the samples was determined via GC-FID analysis and the water content via Karl Fischer titration. Further details are available in Swanepoel and Schwarz (2017).

VLE and VLLE experimental methodology

The VLE and VLLE data were obtained using a modified PilotDist VLE100 dynamic recirculating vapour liquid equilibrium still (Pienaar et al., 2013). The unit was modified through the addition of an ultrasonic homogenizer, which allows for the measurement of VLLE data. The same procedure was followed for obtaining VLLE data as for VLE data, with the additional steps of using the ultrasonic homogenizer to ensure the organic and aqueous phases are sufficiently emulsified and allowing the liquid sample to separate for a minimum of two hours before drawing samples from each phase for VLLE measurements. The compositions of the samples were determined using GC analysis with use of thermal conductivity detection.

Results and Discussion

The results for the LLE data show that the presence of the second alcohol has a noticeable effect of the LLE in the quaternary systems and needs to be considered, in terms of both alcohol recovery to the organic phase and the size of the immiscible region. The presence of ethanol reduced the concentration of 2-propanol that reports to the organic phase while the presence of 2-propanol increases the concentration of ethanol that reports to the organic phases. The interaction of ethanol with 2-propanol and vice versa therefore significantly influences the phase behaviour of the system.

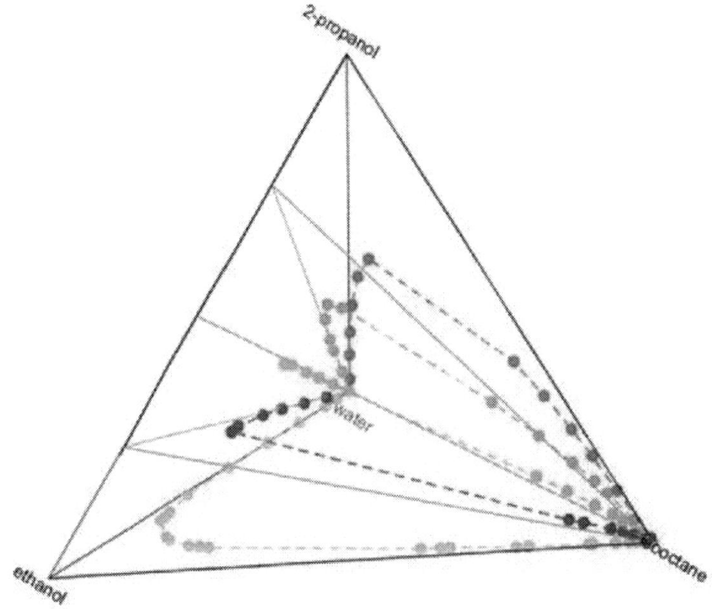

Figure 1. Quaternary diagram representing the liquid-liquid phase equilibrium data for the quaternary system of water + ethanol + 2-propanol + isooctane at 298K and atmospheric pressure, for three 2-propanol: ethanol ratios

Likewise, the effect of the presence of a second alcohol should also be considered for the VLE and VLLE of the quaternary system. The effect of the presence of a second alcohol can be studied in this quaternary water + ethanol + 2-propanol + isooctane system but the insights gained can be applied to systems of other alcohols and/or entrainers.

References

Font, A., Asensi, J. C., Ruiz, F. & Gomis, V., 2003. Application of Isooctane to the Dehydration of Ethanol. Design of a Column Sequence To Obtain Absolute Ethanol by Heterogeneous Azeotropic Distillation. *Industrial and Engineering Chemistry Research,* Volume 42, pp. 140-144.

Font, A., Asensi, J. C., Ruiz, F. & Gomis, V., 2004. Isobaric Vapor-Liquid and Vapor-Liquid-Liquid Equilibria Data for the System Water + Isopropanol + Isooctane. *Journal of Chemical and Engineering Data,* Volume 49, pp. 765-767.

Pienaar, C. Schwarz, C.E., Knoetze, J.H., Burger, A.J., 2013. Vapor-Liquid-Liquid Equilibria Measurements for the Dehydration of Ethanol, Isopropanol, and n-Propanol via Azeotropic Distillation Using DIPE and Isooctane as Entrainers, *Journal of Chemical and Engineering Data*, Volume 58, pp 537-550.

Swanepoel, R. M. & Schwarz, C. E., 2017. Influence of Temperature on the Liqiud-Liquid Phase Equilibria of Ternary (Water + Alcohol + Entrainer) Systems. *Journal of Chemical and Engineering Data,* 62, pp. 2740-2754.

575269 The Influence of Ammonia on Flame Characteristics and Soot Formation in Ammonia/Methane Diffusion Flames

Because of narrow stability limits for the efficient burning of ammonia (NH_3), strategies for enhancing the stability of NH_3 combustion processes have been pursued, such as through co-firing with hydrocarbons. Under these circumstances soot formation becomes possible, and the influence of NH_3 on soot emissions in these technologies becomes relevant. However, there is a lack of studies that have analyzed the effect of NH_3 on soot formation pathways and how it ultimately impacts particulate matter emissions. In order to begin answering these questions, we measured the flame structure characteristics and relative soot volume fraction distributions in NH_3-CH_4 co-flow diffusion flames with varying levels of NH_3-blending. We also compared these measurements to those in N_2-diluted CH_4 flames for reference. Center-line temperature measurements and flame images were taken to determine flame heights and lift-off heights, respectively, and relative maximum soot volume fractions in flames with varying ratios of NH_3 and CH_4 were measured using line-of-sight spectral radiance. Computational 2D flame simulations were also employed in order to estimate flame lengths, lift-off heights, and distributions of species relevant to soot formation. Experimental and computations results are compared against each other in order to assess the accuracy of the underlying chemical mechanism and begin explaining the soot formation pathways affected by the presence of NH_3.

575399 Evaluation of Para- and Ortho-Xylene Adsorption on Beta Zeolite Using the Headspace Chromatography

Xylenes isomers are aromatic compounds obtained from the petroleum and have a large industrial importance. Separation of xylene isomers is a key issue in the petrochemical industry. In this study, the headspace chromatography method was used to evaluate the selectivity of adsorption of o-xylene and p-xylene on Beta zeolite (Si/Al = 25 and crystal diameter = 6 μm). This technique consists of evaluate the vapor phase in equilibrium with the liquid and adsorbed phases in a sealed cell. This method is very useful for the selection of adsorbents, as well as in the investigation of selectivity variation at different conditions. The experiments were carried out at different temperatures and molar ratio of p-xylene/o-xylene. Results are comparable to other more costly and time-consuming techniques. Para-xylene selectivity was observed for the Beta zeolite sample ($a_{px/ox} > 2$).

558742 Study of the Influence of Antifoaming Additives on the Crude OIL/Air Interface By Rheology

Each crude oil reservoir has unique conditions of temperature and pressure, where the crude oil has different amounts of dissolved gas in its constitution. During production, as the fluid rises in the pipes, the pressure and temperature decrease, causing the previously dissolved gas to be liberated. This activity, along with the agitation that occurs during the extraction process, leads to the formation of foams.

The foams generated during oil production are complex systems, making it hard to understand what occurs in the foam layers, due to the complexity of their continuous phase. Some studies have shown that the main species that stabilize foams are asphaltenes, because they reduce the surface tension of the bubbles due to their interfacial activity. The interfacial layer of the foam becomes viscoelastic due to the presence of these amphiphilic species, which have superficial action and can be adsorbed through the dispersed or continuous phase. Therefore, rheological analysis of this interface is necessary to understand its characteristics and stability, since the lifetime of this foam is closely associated with these conditions.

Interfacial rheology also provides information on the functional interplay of the deformation of an interface, the forces acting on it and the resulting flows in the adjacent phases of the fluid. This can be accomplished by dilatation and shear methods.

The aim of this work was to study the crude oil/air interface of different crude oil samples and different antifoaming products using the BiCone accessory, which is used in the shear method.

The tests of the foam formation in crude oil and the rheology of the crude oil/air interface were performed with or without the presence of antifoaming additives, using two samples of crude oil from Brazilian wells (P1 - 19 °API and P2 - 26 °API).

In this study, commercial silicone-based antifoaming agents and formulations composed of a nonionic surfactant based on poly(ethylene oxide) were used.

The tests of the foam formation in the two crude oil samples were carried out according to the method described by Fraga[1] and Rezende[2]. The tests were performed using a Fann High Temperature Aging Cell® and a Fann 704Es roller oven. All the tests were performed in duplicate and the means of the data were used to plot the graphs. The tests in the presence of the antifoaming additives were carried out at concentrations of 20 ppm or 100 ppm.

The rheological properties of the interfacial films of the crude oil/air systems were determined with a Haake MARS 60 rotational rheometer using the BiCone accessory. The main objective of this study was to observe the influence of the antifoaming formulations on the values G' and G" (elastic modulus and viscous modulus, respectively) of the interfacial films, since these properties are very important to understand the antifoaming action mechanism at the crude oil/air interfacial layer.

In all the tests, the crude oil samples, with or without the antifoaming formulations, were left in the accessory for 17 hours (aging time), to assure adsorption of the asphaltenic fractions at the crude oil/air interface. We performed oscillatory shear tests with variation to time but constant shear rate of 10% and frequency of 1 Hz.

The results indicate there was a reduction of the G'/G" ratio of the interfacial film formed by sample P1 and P2 with addition of 20 ppm of the commercial antifoaming agents in relation to the pure crude oil. This interfacial rheology result corroborates the results of the tests of foam formation efficiency, where it was possible to observe that the height of the foam formed by that crude oil sample with the addition of additives was considerably lower than that of the pure crude oil.

The study reported here showed that rheology measurements of the crude oil/air interface utilizing the

BiCone accessory can be correlated with foam formation tests of crude oil samples.

References:

1. Fraga, A. K.; Rezende, D. A.; Santos, R. F.; Mansur, C. R. E. Braz. *Petrol. Gas*. **2011**, 5(1), 25-33.

2. Rezende, D. A. Bittencourt, R. R., Mansur, C. R. E. *Petrol. Sci. Eng*. **2011**, 76, 172-177.

567286 Experimental and Theoretical Study of Viscosity Behavior of Sulfonated Partially Hydrolyzed Polyacrylamides (S-HPAM) with Long-Term Thermal Stability

With the application of polymer flooding in enhanced oil recovery, great efforts have been made in improving polyacrylamide performance with high temperature and high salinity. Here in this paper, our effort is focused on developing a novel HPAM copolymers with long term thermal stability and theoretical simulations were done to investigate the mechanism of its viscosity behavior.

By aqueous solution polymerization, temperature and salt-resistant groups were introduced into the traditional HPAM structures and a modified high temperature-resistant sulfonated polyacrylamide (termed as S-HPAM) was developed. The viscosity behavior was studied with respect to temperature, salinity, rejection rate and aging with time. Besides, molecular dynamics (MD) simulations using Gromacs software were performed to study the influence of temperature and inorganic salts on the viscosity behavior of the designed S-HPAM in solution at the atomic and molecular level.

The designed S-HPAM exhibited a relatively high temperature and salt tolerance. Aged at 95°C in oilfield water for 180 days, the viscosity retention rate of S-HPAM demonstrated excellent thermal stability with a slight decrease in viscosity, which meant great potential in long-term thermal stability. Besides, S-HPAM showed better shear-resistant property, as the injection rate increased to 600 mL/h, higher viscosity retention rate were observed. MD simulation results indicated that the introduced temperature and salt-resistant groups brought changes of the charged groups, flexibility and intermolecular non-bonded interaction, which endowed S-HPAM with long-term thermal stability. The improved temperature resistance performance may be contributed to three factors. First, the introduced sulfonic acid group weaken the electrostatic shielding effect of salt ions on molecular chains. Second, the large side chains introduced increase the shear resistance, and the steric hindrance of side chains enhanced the hydrolysis resistance of adjacent amino groups. Furthermore, a higher non-bond interaction energy of S-HPAM lead to stronger resistance to temperature and salt ions.

The modified S-HPAM exhibited improved stability under high temperature and salinity environment and its long-term thermal stability showed great potential in the application of polymer flooding. Besides, MD simulations suggested an effective way to improve temperature resistance by tuning charged groups, side chains and non-bond interaction of polymers.

582034 Extractive Denitrogenation and Desulfurization of Crude Oil Using Deep Eutectic Solvents

Hydrotreatment is the established method in the industry for removal of aromatics and heteroaromatics from crude oil. Conventionally, hydrotreatment is conducted at relatively high pressure (~ 3.5 - 7 MPa) and temperature (~ 573 – 673 K) resulting in an expensive and energy-consuming process. Moreover, it suffers from high hydrogen consumption and low removal efficiency for some heteroaromatics (i.e. S- and N-containing aromatic) due to steric hindrance. Aiming to reduce the energy requirements, hydrogen consumption and improve the efficiency of this separation, liquid-liquid extraction is proposed. The main objective of this work was to selectively remove the sulfur- and nitrogen-containing aromatics from crude oil using a highly selective solvent as an extractive agent. This allows the removal of (S- and N-containing) heteroaromatic compounds from crude oil at milder conditions (near atmospheric) and without any hydrogen consumption, making the process much more energy- and cost-effective. Deep eutectic solvents (DESs) are proposed as an extractive agent for this application. DESs are analogus to ionic liquids in terms of low volatility and tuneability of physiochemical properties, however, being easier to make from low-cost starting material and more environmentally benign.

Building on the literature finding, phosphonium-based DESs were selected for the removal of S- and N-containing aromatic from crude oil "represented by *n*-alkanes". Next, solubility measurements of S- and N-containing aromatics and *n*-alkanes in DESs were carried out to examine the suitability of the DESs for this extraction. Then, the single stage liquid-liquid equilibrium data of {*n*-alkane + heteroaromatic + DES} systems were determined at 298 K and atmospheric pressure. The experimental LLE data were correlated using the nonrandom two-liquid thermodynamic model. Also, the solute distribution ratios and the selectivities were calculated from the experimental LLE data and compared to the relevant literature. It was found that the DESs are potential solvents for extractive desulfurization processes.

582489 Processing Brazilian Pre-salt Crudes

Processing Brazilian Pre-salt crudes has already become a reality for PETROBRAS and Brazil. Just 10 years after startup, Pre-salt oil reached half the volume produced by PETROBRAS and is expected to keep on growing. These oils present intermediate API gravity, low naphthenic acidity, low sulphur and intermediate nitrogen contents. Their paraffinicity favors FCC conversion, naphtha quality, higher cetane and higher API gravity diesel; and its low sulphur resid fractions boost bunker production. Parallel to the expansion in Pre-salt production, PETROBRAS has gained experience in refining these crudes in several domestic refineries and has therefore customized a set of operational best practices and technical guidelines to achieve the best results for their processing. This poster will highlight technical challenges faced and successful solutions adopted by PETROBRAS Refining for processing Pre-salt Crudes.

571766 Electrochemical Characterization of $SrTi_{0.3}Fe_{0.7}O_{3-\Delta}$-GDC Composite Anodes for Solid Oxide Fuel Cell

In this work, $SrTi_{0.3}Fe_{0.7}O_{3-d}$-GDC (STF-GDC; GDC, $Gd_{0.1}Ce_{0.9}O_{1.95}$) composite is adopted as anode materials to systematically investigate the effect of GDC addition on the anode stability and electrochemical property. $La_{0.8}Sr_{0.2}Ga_{0.8}Mg_{0.2}O_{3-d}$ (LSGM) electrolyte supported single cells with STF-GDC anode and LSCF-GDC/LSCF (LSCF, $La_{0.4}Sr_{0.6}Co_{0.2}Fe_{0.8}O_{3-d}$) cathode are prepared via screen printing and high temperature calcination in air. The weight percentages of GDC in composite anodes is 10%, 20%, 30% and 40%. $La_{0.4}Ce_{0.6}O_{2-d}$ is used as barrier layer between anode and LSGM electrolyte. Electrochemical Impedance Spectrum (EIS) and single cells output performance is tested under different anode hydrogen partial pressure. A typical equivalent circuit model is used to fit the EIS data. Fitting results show three responses and only high frequency response is nearly no change with decreasing of hydrogen partial pressure so it could be attributed to the cathode process. Moreover, it is observed that with the amount of GDC increasing, current density generated at 0.7 V decreases more quickly with decreasing of hydrogen partial pressure. At last, the van Der Pauw DC conductivity test is conducted to characterize the conducting property of the STF-GDC composite anodes.

Ceria-Lanthanide-Silver Oxides for Solar Thermochemical Conversion of CO_2

(Principal Author)
Gorakshnath Takalkar
Department of Chemical Engineering
College of Engineering
Qatar University, Doha, Qatar

(Co-Authors)
Rahul Bhosale
Department of Chemical Engineering
College of Engineering
Qatar University, Doha, Qatar
rahul.bhosale@qu.edu.qa

Fares AlMomani
Department of Chemical Engineering
College of Engineering
Qatar University, Doha, Qatar

Suliman Rashid
Department of Chemical Engineering
College of Engineering
Qatar University, Doha, Qatar

Prepared for Presentation at
American Institute of Chemical Engineers
2019 Annual Meeting
Orlando, FL, USA
November 10 – 15, 2019

Ceria-Lanthanide-Silver Oxides for Solar Thermochemical Conversion of CO_2

Gorakshnath Takalkar, Rahul R. Bhosale*, Fares AlMomani, Suliman Rashid

Department of Chemical Engineering
College of Engineering
Qatar University, Doha, Qatar

Keywords: Ceria, Silver, Lanthanide, CO_2 splitting, Co-Precipitation

Abstract

This paper reports the thermochemical splitting of CO_2 using $Ce_{0.9}Ln_{0.05}Ag_{0.05}O_{2-\delta}$ materials, which are prepared by using the co-precipitation method. The phase composition and the morphology of the derived materials was analyzed using powder X-ray diffractometer and scanning electron microscope. The thermal reduction and CO2 splitting reactions were conducted by using a high temperature TGA operated in the range of 1000 to 1400°C.

1. Introduction

The increasing level of CO_2 in the earth's atmosphere and requirement in the energy needed to drive the daily life on earth, it is highly essential to apply CO_2 capture and conversion technologies on a large scale[1-3]. Among the multiple metal oxides investigated for solar driven thermochemical splitting of water and carbon dioxide, the ceria based doped oxides are considered as one of the promising [4-52]. Recently, we have published an article in which the fuel production capacity of the ceria-transition metal oxides has been improved by incorporating a small amount of silver. By considering this study as our basis, in this investigation, we have attempted to explore the thermal reduction and CO_2 splitting capabilities of the ceria-lanthanide-silver (CLS) oxides. The reactions associated with CLS oxide based solar driven thermochemical CO_2 splitting process are listed below.

$$Ce_xLn_{1-x-0.05}Ag_{0.05}O_2 \rightarrow Ce_xLn_{1-x-0,05}Ag_{0.05}O_{2-\delta} + O_2(g) \quad (1)$$

$$Ce_xLn_{1-x-0.05}Ag_{0.05}O_{2-\delta} + CO_2 \rightarrow Ce_xZn_{1-x}O_2 + CO(g) \quad (2)$$

2. Experimental

2.1 Synthesis of CLS oxides

For the preparation of the CLS oxides, the nitrate precursors and the precipitating agent (aqueous ammonia hydroxide) were procured from Sigma Aldrich. All the samples were prepared using the deionized water produced from ultrapure Type 1 from Direct –Q system,

Millipore, France. The gases utilized for the TGA experiments were ordered in cylinder from Buzwair Scientific and Technical Gases supplier, Qatar.

As the first step towards synthesis of CLS oxides, the calculated amounts of the metal precursors were dissolved in the deionized water. Once the dissolution is achieved, the pH of the solution was maintained close to 10 by adding the aqueous ammonia hydroxide in a dropwise manner. The mixture was allowed to stir for 40 to 50 hours (with a maintained pH equal to 10). Once the stirring time was completed, the precipitate was allowed to settle down (overnight). The settled precipitate was washed for multiple times by using the deionized water. The solids collected after filtration were dried, crushed and annealed in presence of air up to 800 °C for 4 hours. The obtained calcined powder was stored and then analyzed further towards characterization and TGA.

2.2 TGA Experiments

The redox reactivity of the CLS oxides towards CO_2 conversion was scrutinized by performing the thermochemical experiments in a thermogravimetric analyzer (TAG) which is shown in Fig. 1.

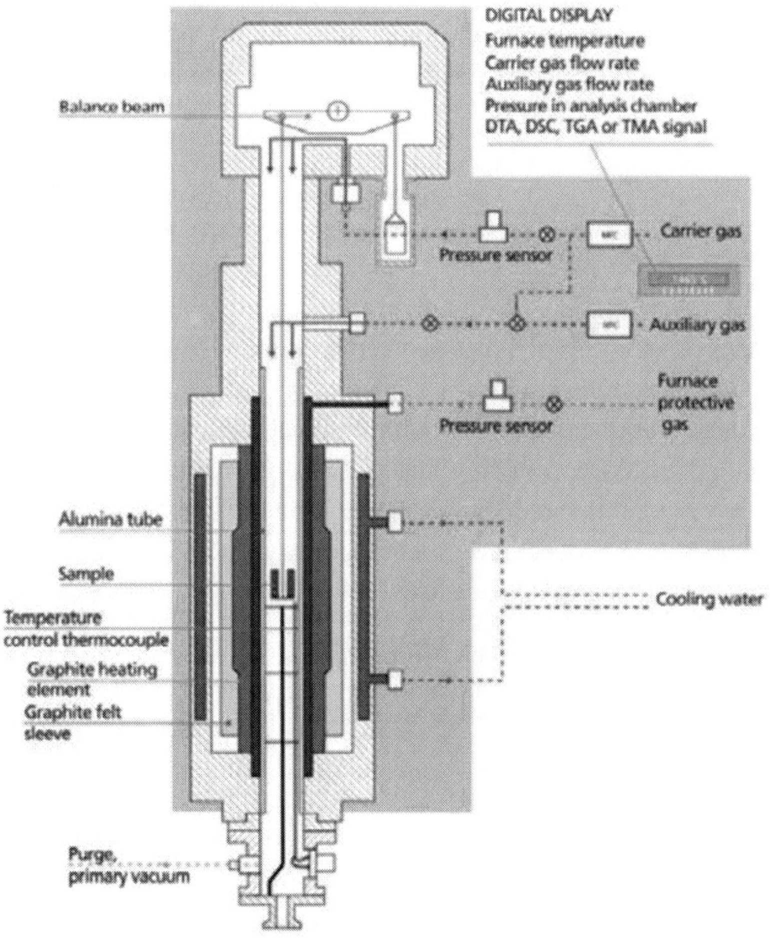

Figure 1. TGA used for the splitting of CO_2 using CLS oxides.

The typical TGA profile obtained for one the CLS oxide is shows in Fig. 2. Selection of the operating conditions was done based on the previous studies. Multiple thermochemical cycles were employed, the amounts of O2 released, and CO produced by each CLS oxide was estimated. The buoyancy effect was eliminated by subtracting the blank run from the actual experiments.

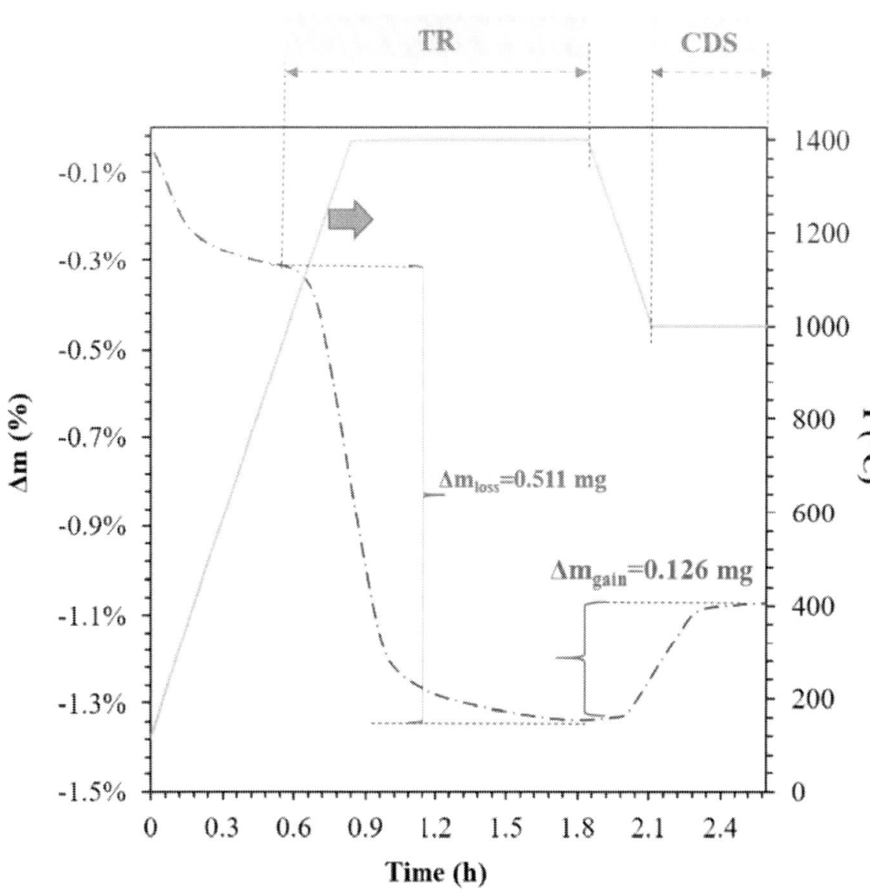

Figure 2. Typical TGA profile obtained for one of the CLS oxide.

3. Results and Discussion

The CLS oxides were synthesized using the co-ppt method and the obtained powder was analyzed using multiple analytical techniques. After the successful synthesis and characterization, the CLS oxides were examined towards thermochemical CO2 splitting cycles by using a high temperature TGA. The TGA profiles obtained provides initial insight into the redox reactivity of CLS oxides towards thermochemical conversion of CO2. For the employed experimental condition, all the materials were observed to be capable of thermal reduction and re-oxidation via CO2 splitting reaction. Overall, the La-doped ceria-Ag (CLaS) oxide was observed to be the most promising in terms of amounts of O2 released and CO produced.

4. Conclusions

Various combinations of Ce-Ln-Ag based oxides were derived by using the co-precipitation of hydroxide method. All these materials were characterized by using multiple analytical techniques. The derived materials were further tested towards CO2 splitting experiment using TGA. The obtained results indicate that most of the materials were capable of producing higher amounts of CO as compared to ceria.

5. Acknowledgments

This publication was made possible by the NPRP grant (NPRP8-370-2–154) from the Qatar National Research Fund (a member of Qatar Foundation). The statements made herein are solely the responsibility of author(s).

References

[1] R. R. Bhosale, V. V. Mahajani, Kinetics of Absorption of Carbon Dioxide in Aqueous Solution of Ethylaminoethanol Modified with N-methyl-2-pyrolidone, Separation Science and Technology 48 (2013) 2324-2337.

[2] R. R. Bhosale, V. V. Mahajani, Kinetics of thermal degradation of renewably prepared amines useful for flue gas treatment, Journal of Renewable and Sustainable Energy 5 (2013) 063110.

[3] R. R. Bhosale, A. Kumar, F. AlMomani, U. Ghosh, A. AlNouss, J. Scheffe, R. B. Gupta, CO_2 Capture Using Aqueous Potassium Carbonate Promoted by Ethylaminoethanol: A Kinetic Study, Industrial and Engineering Chemistry Research 55 (2016) 5238-5246.

[4] J.R. Scheffe, A. Steinfeld, Oxygen exchange materials for solar thermochemical splitting of H_2O and CO_2: a review, Materials Today 17 (2014) 341–348.

[5] R. R. Bhosale, G. D. Takalkar, P. N. Sutar, A. Kumar, F. AlMomani, M. Khraisheh, A decade of ceria based solar thermochemical H_2O/CO_2 splitting cycle, International Journal of Hydrogen Energy 44 (2018) 34-60.

[6] R. R. Bhosale, I. Alxneit, L.J.P. Van Den Broeke, A. Kumar, M. Jilani, S.S. Gharbia, J. Folady, D. Z. Dardor, Sol-gel synthesis of nanocrystalline Ni-ferrite and Co-ferrite redox materials for thermochemical production of solar fuels, Materials Research Society Symposium Proceedings 1675 (2014) 203-208.

[7] D. Yadav, R. Banerjee, A review of solar thermochemical processes, Renewable and Sustainable Energy Reviews 54 (2016) 497 – 532.

[8] R. R. Bhosale, A. Kumar, F. AlMomani, R. B. Gupta, Solar thermochemical $ZnO/ZnSO_4$ water splitting cycle for hydrogen production, International Journal of Hydrogen Energy 43 (2017) 23474-23483.

[9] P.G. Loutzenhiser, A. Steinfeld, Solar syngas production from CO2 and H2O in a two-step thermochemical cycle via Zn/ZnO redox reactions: thermodynamic cycle analysis, International Journal of Hydrogen Energy 36 (2011) 12141–12147.
[10] R. R. Bhosale, A. Kumar, P. Sutar, Thermodynamic analysis of solar driven SnO_2/SnO based thermochemical water splitting cycle, Energy Conversion & Management 135 (2017) 226-235.
[11] R. R. Bhosale, P. Sutar, A. Kumar, F. AlMomani, M. H. Ali, U. Ghosh, S. AlMuhtaseb, M. Khraisheh, Solar hydrogen production via erbium oxide based thermochemical water splitting cycle, Journal of Renewable and Sustainable Energy 8 (2016) 034702.
[12] A. Stamatiou, P. Loutzenhiser, A. Steinfeld, Solar syngas production from H_2O and CO_2 via two-step thermochemical cycles based on Zn/ZnO and FeO/Fe_3O_4 redox reactions: Kinetic analysis, Energy & Fuels 24 (2010) 2716-2722.
[13] R.R. Bhosale, A. Kumar, F. AlMomani, Solar Thermochemical Hydrogen Production via Terbium Oxide Based Redox Reactions, International Journal of Photoenergy 2016 (2016) 9727895.
[14] P. Loutzenhiser, M. Gálvez, I. Hischier, A. Stamatiou, A. Frei, A. Steinfeld, CO_2 splitting via two-step solar thermochemical cycles with Zn/ZnO and FeO/Fe_3O_4 redox reactions II: Kinetic analysis, Energy & Fuels 23 (2009) 2832-2839.
[15] R. Bhosale, R. Shende, J. Puszynski, Sol-gel derived $NiFe_2O_4$ modified with ZrO_2 for hydrogen generation from solar thermochemical water-splitting reaction, Materials Research Society Symposium Proceedings 1387 (2011) 34 – 39.
[16] P. Charvin, S. Abanades, F. Lemont, G. Flamant, Experimental study of SnO2/SnO/Sn thermochemical systems for solar production of hydrogen, AIChE Journal 54 (2008) 2759-2767.
[17] R. V. Shende, J. Puszynski, M. Opoku, R. Bhosale, Synthesis of novel ferrite foam material for water-splitting application, Proceedings of the 2009 NSTI Nanotechnology Conference and Expo., 1 (2009) 201 – 204.
[18] C.C. Agrafiotis, C. Pagkoura, A. Zygogianni, G. Karagiannakis, M. Kostoglou, A.G. Konstandopoulos, Hydrogen production via solar-aided water splitting thermochemical cycles: Combustion synthesis and preliminary evaluation of spinel redox-pair materials, International Journal Hydrogen Energy 37 (2012) 8964-8980.
[19] R.R. Bhosale, A. Kumar, F. AlMomani, U. Ghosh, M. S. Anis, K. Kakosimos, R. V. Shende, M. A. Rosen, Solar hydrogen production via a samarium oxide-based thermochemical water splitting cycle, Energies 9 (2016).
[20] M. Roeb, N. Gathmann, M. Neises, C. Sattler, R. Pitz-Paal, Thermodynamic analysis of two-step solar water splitting with mixed iron oxides, International Journal of Energy Research 33 (2009) 893-902.
[21] R. R. Bhosale, A. Kumar, F. AlMomani, U. Ghosh, P. Sutar, G. Takalkar, A. Ashok, I. Alxneit, Effectiveness of Ni incorporation in iron oxide crystal structure towards thermochemical CO_2 splitting reaction, Ceramics International 43 (2017) 5150-5155.

[22] J.R. Scheffe, J. Li, A.W. Weimer, A spinel ferrite/hercynite water-splitting redox cycle, International Journal of Hydrogen Energy 35 (2010) 3333-3340.

[23] R. R. Bhosale, A. Kumar, F. AlMomani, U. Ghosh, D. Dardor, Z. Bouabidi, M. Ali, S. Yousefi, A. AlNouss, M. S. Anis, M. H. Usmani, M. H. Ali, R. S. Azzam, A. Banu, Solar co-production of samarium and syngas via methanothermal reduction of samarium sesquioxide, Energy Conversion and Management 112 (2016) 413 – 422.

[24] J.R. Scheffe, A. Steinfeld, Thermodynamic analysis of cerium-based oxides for solar thermochemical fuel production, Energy & Fuels 26 (2012) 1928-1936.

[25] R. R. Bhosale, R. V. Shende, J. Puszynski, Sol-gel synthesis of ferrite foam materials for H2 generation from water-splitting reaction, Proceedings of the 2010 NSTI Nanotechnology Conference and Expo., 1 (2010) 368 – 371.

[26] W.C. Chueh, C. Falter, M. Abbott, D. Scipio, P. Furler, S.M. Haile, A. Steinfeld, High-flux solar-driven thermochemical dissociation of CO_2 and H_2O using nonstoichiometric ceria, Science 330 (2010) 1797-1801.

[27] R. R. Bhosale, R. V. Shende, J. Puszynski, Synthesis of nano-sized ferrite materials for H_2 generation from high temperature water-splitting reaction, Proceedings of the 2011 NSTI Nanotechnology Conference and Expo., 1 (2010) 237 – 240.

[28] J.R. Scheffe, R. Jacot, G.R. Patzke, A. Steinfeld, Synthesis, Characterization, and Thermochemical Redox Performance of Hf+4 , Zr+4 , and Sc+3 Doped Ceria for Splitting CO2, The Journal of Physical Chemistry C 117 (2013) 24104-24114.

[29] R. R. Bhosale, R. V. Shende, J. Puszynski, H2 generation from thermochemical water-splitting using sol-gel synthesized ferrites (MxFeyOz, M = Ni, Zn, Sn, Co, Mn, Ce), Proceedings of the 2010 AIChE Annual Meeting (2010).

[30] J.R. Scheffe, D. Weibel, A. Steinfeld, Lanthanum–strontium–manganese perovskites as redox materials for solar thermochemical splitting of H_2O and CO_2, Energy & Fuels 27 (2013) 4250-4257.

[31] R. R. Bhosale, R. V. Shende, M. K. Opoku, J. Puszynski, H_2 generation from water-splitting using novel ferrite foam materials, Proceedings of the 2009 AIChE Annual Meeting (2009).

[32] A.H. McDaniel, E.C. Miller, D. Arifin, A. Ambrosini, E.N. Coker, R. O'Hayre, W.C. Chueh, J. Tong, Sr-and Mn-doped $LaAlO_{3-\delta}$ for solar thermochemical H_2 and CO production, Energy & Environmental Science 6 (2013) 2424-2428.

[33] R. R. Bhosale, A. Kumar, A. Ashok, P. Sutar, G. D. Takalkar, M. Khraisheh, F. AlMomani, U. Ghosh, La-Based Perovskites as Oxygen-Exchange Redox Materials for Solar Syngas Production, MRS Advances 2 (2017) 3365-3370.

[34] M. Gálvez, R. Jacot, J. Scheffe, T. Cooper, G. Patzke, A. Steinfeld, Physico-chemical changes in Ca, Sr and Al-doped La–Mn–O perovskites upon thermochemical splitting of CO_2 via redox cycling, Physical Chemistry Chemical Physics 17 (2015) 6629-6634.

[35] R. R. Bhosale, X. Pasala, S. Yelkanti, J. A. Puszynski, R. V. Shende, Solar thermochemical water-splitting for H_2 generation using sol-gel derived ferrite nanomaterials, Proceedings of the 2009 NSTI Nanotechnology Conference and Expo., (2012) 555 – 558.

[36] F. A. Costa Oliveira, M. A. Barreiros, S. Abanades, A. P. F. Caetano, R. M. Novais, R. C. Pullar, Solar thermochemical CO2 splitting using cork-templated ceria ecoceramics, Journal of CO_2 Utilization 26 (2018) 552-563.
[37] R.R. Bhosale, A. Kumar, F. AlMomani, U. Ghosh, S. Al-Muhtaseb, R. Gupta, I. Alxneit, Assessment of $Ce_xZr_yHf_zO_2$ based oxides as potential solar thermochemical CO_2 splitting materials, Ceramics International 42 (2016) 9354-9362.
[38] M. Hoes, C. Muhich, R. Jacot, G. R. Patcke, A. Steinfeld, Thermodynamics of paired charge-compensating doped ceria with superior redox performance for solar thermochemical splitting of H_2O and CO_2, Journal of Material Chemistry A 5 (2017) 19476-19484.
[39] R.R. Bhosale, A. Kumar, F. AlMomani, I. Alxneit, Sol–gel derived CeO_2–Fe_2O_3 nanoparticles: Synthesis, characterization and solar thermochemical application, Ceramics International 42 (2016) 6728-6737.
[40] C. N. R. Rao, S. Dey, Generation of H_2 and CO by solar thermochemical splitting of H_2O and CO_2 by employing metal oxides, Journal of Solid State Chemistry 242 (2016) 107-115.
[41] R. R. Bhosale. A. Kumar, F. AlMomani, I. Alxneit, Propylene oxide assisted sol–gel synthesis of zinc ferrite nanoparticles for solar fuel production, Ceramics International 42 (2016) 2431-2438.
[42] Y. Hao, C. Yang, S. Haile, Ceria–Zirconia Solid Solutions ($Ce_{1-x}Zr_xO_{2-\delta}$, $x \leq 0.2$) for Solar Thermochemical Water Splitting: A Thermodynamic Study, Chemistry of Materials 26 (2014) 6073-6082.
[43] R. R. Bhosale. A. Kumar, F. AlMomani, U. Ghosh, M. Khraisheh, A comparative thermodynamic analysis of samarium and erbium oxide based solar thermochemical water splitting cycle, International Journal of Hydrogen Energy 42 (2017) 23416-23426.
[44] T. Cooper. J. Scheffe, M. E. Galvex, R. Jacot, G. Patzke, A. Steinfeld, Lanthanum Manganite Perovskites with Ca/Sr A-site and Al B-site Doping as Effective Oxygen Exchange Materials for Solar Thermochemical Fuel Production, Energy Technology 3 (2015) 1130 – 1142.
[45] R. R. Bhosale. Thermodynamic analysis of Ni-ferrite based solar thermochemical H2O splitting cycle for H_2 production, International Journal of Hydrogen Energy 44 (2019) 61-71.
[46] C. Muhich, A. Steinfeld, Principles of doping ceria for the solar thermochemical redox splitting of H_2O and CO_2, Journal of Materials Chemistry A 5 (2017) 15578-15590.
[47] R. R. Bhosale, G. D. Takalkar, Nanostructured co-precipitated $Ce_{0.9}Ln_{0.1}O_2$ (Ln = La, Pr, Sm, Nd, Gd, Tb, Dy, or Er) for thermochemical conversion of CO_2, Ceramics International 44 (2018) 16688–16697.
[48] D. Arifin, A. W. Weimer, Kinetics and mechanism of solar-thermochemical H_2 and CO production by oxidation of reduced CeO_2, Solar Energy 160 (2018) 178-185.
[49] R. R. Bhosale, Synthesis and characterization of nanocrystalline $CoFe_2O_4$ – zirconia via propylene oxide aided sol-gel method, Ceramics International 44 (2018) 191-197.
[50] R. R. Bhosale, Thermodynamic efficiency analysis of zinc oxide based solar driven thermochemical H_2O splitting cycle: Effect of partial pressure of O_2, thermal reduction and H_2O splitting temperatures, International Journal of Hydrogen Energy 43 (2018) 14915–14924.

[51] R. R. Bhosale, Solar driven two-step CH_4 reforming and H_2O splitting using Al_2O_3 for Co-production of Al, syngas, and H_2, Solar Energy 172 (2018) 232-241.
[52] G.D. Takalkar, R.R. Bhosale, A. Kumar, F. AlMomani, M. Khraisheh, R.A. Shakoor, R.B. Gupta, Transition metal doped ceria for solar thermochemical fuel production, Solar Energy (2018) Article In-Press.

576882 Advanced Optical Fiber Probe Technique to Measure Local Liquid Saturation and Velocity within Various Commercial Catalyst Shapes in Pilot Plant Trickle Bed Reactors (TBRs)

Advanced Two-Tip Optical Fiber Probe Technique to Measure Local Gas and Liquid Saturations and Velocities within Various Commercial Catalyst Shapes in Pilot Plant Trickle Bed Reactors (TBRs).

Mohammed Al-Ani[a], Muthanna H. Al-Dahhan[ab]

a Chemical and Biochemical Engineering Department, Missouri University of Science and Technology, Rolla MO 65409, USA

b Mining and Nuclear Engineering Department, Missouri University of Science and Technology, Rolla, MO 65409, USA

Email: aldahhanm@mat.edu

Trickle bed reactors (TBRs) have been widely used in multiple commercial processes with different design and operation conditions such as chemical industries (e.g., Petroleum petrochemical, chemical processes, etc…). Trickle bed reactors are packed beds of catalyst particles with concurrent downward of gas and liquid. Broadly, trickle bed reactors are used in various industrial fields due to simplicity in operation under high temperature and high-pressure conditions require for most industrial processes, low energy consumption due to the solids are not suspended. However, the disadvantages of trickle bed reactors are rapid deactivation and the possibility for liquid maldistribution, which may be the reason for formation hot spots and runaway the reactors due to excessive temperature and pressure in the reactor.

Local liquid saturation investigation in multiphase flow rectors can play a significant role on different critical parameters in the reactor (pressure drop, a residence time of the liquid phase, axial dispersion coefficient and heat transfer coefficient, etc.). Local liquid saturation measuring and mapping at different bed height axially and diametrical profile radially with varies operating conditions such as gas and liquid flow rates, bed characteristics (shape and size), and physical properties of fluids flow is essential for design and scale up and scale down engineering to evaluate fundamental information in the reactor.

Advanced two-tip optical fiber probe (OFP) technique was manufactured and developed in Multiphase Flow Reactor Engineering and Application Laboratory (MFReal) in Missouri University of Science and Technology. The main concept for OFP is based on the variation in the light reflection that encountered at the probe tip by gas and liquid in the reactor. Gases revealed higher reflection for the light than liquids. The difference in the light reflection between the gas and liquid enables to measure the local gas and liquid volume fraction locally. OFP has been implemented to study important hydrodynamics in various multiphase flow reactors systems successfully. Also, the high capability was revealed by measuring a diametrical profile along bed height axially. The obtained data by OFP were validated with the data obtained using x-ray digital industrial radiography (DIR) and syringe pump techniques. The results have been indicated a high capability and reliability for OFP to measure dynamic flow parameters (local gas and liquid saturation and velocities).

Moreover, some other parameters such as local gas and liquid velocities and evaluating channeling or bypassing locally along bed height are also capable of being measured by processing different concepts. Successfully, OFP used to study the effects of bed height, gas flow rate, liquid flow rate, and catalyst shape (using different industrial catalyst shapes such as trilobe, quadralobe, cylindrical, and spherical particles) on the liquid saturation and local gas and liquid velocities diametrical profiles studied to evaluate performance in the pilot plant trickle bed reactor.

555773 Refinery Vacuum Ejector Systems: A Process Engineer's Perspective

The largest refinery vacuum systems are on the crude vacuum tower overhead. The vacuum system is a utility system that lowers the vacuum tower pressure to allow more of the feed to the tower to vaporize. They are required to handle a mixture of water vapor, hydrocarbons, and other gases from the vacuum tower. Steam jet ejectors are the dominant vacuum equipment used in these systems. The entire vacuum system includes more than just the ejectors. It includes the ejectors, condensers, and hot-well. How these components interact determines overall system performance. When dealing with vacuum systems designers, operators, and troubleshooters must have a fundamental understanding of how the components work and interact.

Topics covered here include:

- Vacuum process type (wet versus dry/damp systems)
- Load factors
- Basic component operation
- Component interaction
- Common alternate configurations

561057 Influence of Long Time on Stream (TOS) in the Morphological and Structural Characteristics of Iron-Based Catalyst during FTS. A Closer Look at the Deactivation Routes

Iron catalysts is mostly used in the Fischer-Tropsch synthesis (FTS) because of its prize and ubiquitous in nature. Sintering, oxidation and carbon deposition during Fischer-Tropsch synthesis has long been postulated as some major deactivation mechanisms. However, to date there is no consistent picture. This paper presents deactivation modes of a commercial catalysts after running for long TOS by means of oxidation, sintering and graphitisation for iron-based catalyst. Furthermore, it presents results on the deactivation of an industrial Fe/SiO_2 catalyst as obtained by XRD, XPS, FTIR, HRTEM and TGA. These analyses were performed to study the phases of the iron catalyst and the nature of the carbon deposited on the catalyst of the spent industrial Fe/SiO_2 catalyst samples withdrawn from a fixed bed reactor operating under realistic FTS conditions, and it was concluded that sintering can be ruled out as a major deactivation mechanism if extended TOS are used. Finally, these data together with all relevant literature were used to create a common view on the sintering, oxidation behaviour of metallic Fe during FTS. The apparent discrepancies in literature on the sintering behaviour of Iron are most likely due to the lack of direct measurements of the iron clusters and due to the comparison of catalysts with varying crystallites sizes, compared at different reactor partial pressures of hydrogen and water (P_{H2O}/P_{H2}). As such sintering cannot be predicted because of time variations due to fluctuations in the amount of water, CH_4, butane and other hydrocarbons in the reactor.

586722 AMMONEVA system

Ammonia is an important product for the chemical industry (apart from nitrogen fertilizers applications) that is produced from no renewable products. Pig and cow slurry contain relevant concentration of ammonia nitrogen which volatization to the atmosphere must be avoided. In this work we explain the performance of a system to recover ammonia from pig slurry, called "Ammoneva System", based on the evaporation and absorption of the ammonia, that can be implemented in all sort of livestock facility (slurry either digested or not), obtaining high grade ammonium and protecting environment.

566777 Limitations on the Residual Solvent Recovery Imposed By the Gangue Characteristics

Non-aqueous solvent extraction (NAE) of bitumen from oil sands has the potential to replace the existing hot-water extraction process. One of the challenges of NAE is to minimize the residual non-aqueous volatile solvent in the resultant gangue after extraction. The gangue typically contains a mixture of solids with a wide range of sizes, connate water, residual solvent, and residual bitumen. It has been shown that solvent (cyclohexane in this study) evaporation mainly occurs in the initial steady state of drying in which there is a liquid film connectivity from the inside to the surface of the gangue sample. In this work, the initial evaporation rate of cyclohexane from gangue samples with different compositions was studied. Gangue samples with 12% cyclohexane, based on dry solids, and different water, fines, and bitumen contents were prepared. Mass loss of samples was measured at room temperature and ambient pressure and was used to signify the cyclohexane initial evaporation rate. There existed an optimum cyclohexane evaporation rate over a fines content range of 0 – 20%. In particular, it increased from 45.9 g/min.m^2 in the sample with no fines to 53.9 g/min.m^2 in the sample with 10% fines. However, it dropped to 39.8 g/min.m^2 in the sample with 20% fines. There also existed an optimum water concentration (3.7%) at which the cyclohexane evaporation rate was the highest. Residual bitumen content showed the most drastic impact compared to other components with a 30% reduction in the cyclohexane evaporation rate when only 1% bitumen is present in the gangue mixture. It is proposed that the initial flux of evaporation is controlled by the sorptivity of the sample at low contents of fines, water, and bitumen. The presence of an optimum amount of fines and water seems to increase the intrinsic sorptivity of the sample by decreasing the average pore size of the sample. On the other hand, bitumen reduces the sorptivity by drastically increasing solvent's viscosity. Higher amounts of fines and water leads to the blocking of the pores, thereby entrapping cyclohexane, and consequently reducing its evaporation rate significantly.

572863 Estimation of Phase Equilibrium for Water-Heavy Oil System with Structural Information of Heavy Oil Components

Extra heavy oil, such as Canadian oil sand bitumen, Venezuelan Orinoco tar sand, is an important natural resource to deal with the growth in worldwide oil demand. However, transportation from the wellhead to the refinery or shipping facility presents an important problem for the development of extra heavy crude oil reserves. Supercritical water cracking (SCWC) technology is one of heavy oil partial upgrading technologies to produce the pipeline-transportable synthetic crude oil and this technology has been getting attention as a solution for the problem related the transportation of heavy oil. For the process design of SCWC technology, it is important to understand physical properties of heavy oil and phase equilibrium in water-heavy oil system under the high temperature and high pressure condition.

In the last decade, analytical techniques have advanced significantly, for example, high-resolution mass spectrometry, and this had made it possible to obtain molecule-level information about heavy petroleum feedstocks. In particular, the combination of Fourier transform ion cyclotron resonance mass spectrometry (FT-ICR MS) and soft ionization techniques, such as atmospheric pressure photo ionization (APPI), is an excellent tool to obtain structural and compositional information about highly complex mixtures of hydrocarbons. This technique can provide a better understanding of the structure and composition of heavy petroleum fractions, and these structural and compositional information have a potential to improve the accuracy in the estimation of high temperature and pressure phase equilibrium for water-heavy oil system.

In this study, the estimation method of phase equilibrium with Peng-Robinson equation of state (EoS) and the structural information of heavy oil components is proposed. Middle Eastern atmospheric residue (AR) was used as a typical heavy oil sample. To obtain the structural information, Middle Eastern AR was analyzed by FT-ICR MS with a fragmentation method, collision induced dissociation (CID). In this study, we focused on the core structure which is composed of aromatic rings, naphthenic rings and heteroatom-containing functional groups and which is the most important structure because of its effects on physical properties. An estimation method of core structures comprising heavy oil components was developed and it was confirmed that it could assign core structures contained in heavy oil components. In addition, based on the structural information obtained by the combination of CID FT-ICR MS analysis and the core structure estimation, the correlation and prediction of phase equilibrium in water-heavy oil system was performed with Peng-Robinson EoS. The proposed method could provide the better estimation than the previous method without using the structural information.

565156 Recovery of Heavy Hydrocarbons from Indonesian Asphalt Rocks through Solvent Extraction

Recovery of heavy hydrocarbons from Indonesian asphalt rocks through solvent extraction

Guoqiang MA[a], Hong SUI[a,b,c], Lin HE[a,b], Xingang LI[a,b,c]

[a] School of Chemical Engineering and Technology, Tianjin University, Tianjin 300072, China.

[b] Collaborative Innovation Center of Chemical Science and Engineering(Tianjin), 300072, China.

[c] National Engineering Research Centre of Distillation Technology, Tianjin, 300072, China.

Different from the traditional crude oil, the unconventional oil ores (e.g., oil shale, oil sands, asphalt rocks, etc.) consist of heavy/extra-heavy oil, minerals and water. The geological reserves of unconventional oils (tight oil, natural gas liquids, kerogen oil, extra-heavy oil and bitumen) account for about 64% of total petroleum reserves[1]. It is reported that the crude bitumen production in Alberta reached 2.45 million barrels per day in 2016, and the production of crude bitumen in Alberta should be sufficient to meet Canada¡¯s needs of crude oil for the next several centuries[2]. The Indonesian asphalt rocks, with approximately 100 million tons of asphalt reserves, was firstly reported in 1936 after the discovery of the Buton island[3], which is located off the southeast peninsula of Sulawesi in Indonesia. The Indonesian asphalt rocks, with bitumen content up to 50 wt %, are known as oil-wetted ores without the water film between bitumen and mineral solid surface. The currently used commercial method for bitumen recovery from Athabasca oil sands is the hot water based extraction (HWBE). However, this HWBE process is limited to some extent during the industrial application due to some environmental issues, and the HWBE process has a poor bitumen recovery with the oil-wetted ores[4]. Solvent extraction has been considered a promising process with high recovery rates and can also be employed on various types of oil sands, most notably on oil-wetted oil sands.

In this study, the solvent extraction process has been used to recovery the bitumen from Indonesia asphalt rocks. Both the bitumen recovery, particle sedimentation and the suspended particle flocculation have been investigated. A multi-staged solvent extraction process has been used to recovery the heavy hydrocarbons from Indonesian asphalt rocks with four typical solvents (i.e., toluene, n-heptane, n-hexane, and cyclohexane). Results show that although solvent type-dependent, the solvent extraction gives an extremely high recovery (up to 98%) of heavy hydrocarbons compared with that of HWBE process without water being involved. After the extraction, a systematical particle sedimentation test has been conducted in the non-aqueous phase. The equilibrium height of settled solids is found to be higher in n-hexane and n-heptane systems than those in toluene and cyclohexane systems at given conditions, which is mainly contributed to the higher bitumen components (e.g., asphaltenes) attached on the solid surfaces. A mathematical model has been proposed to describe the carbonate particle settling behaviors for the interfaces in different solvent extraction systems. For the suspended particle in the bitumen solution, different polymer flocculants have been applied, resulting a flocculation rate up to 95%.

These findings provide the application of solvent extraction in recovery the heavy hydrocarbons from Indonesian asphalt rocks. It is believed that the solvent extraction process would be a promising method for the recovery of unconventional oil from the ores.

References:

1. International Energy Agency. *World Energy Outlook 2018*; 2018.

2. Alberta Energy Regulator. *ST98: 2017 Alberta¡¯s Energy Reserves & Supply/Demand Outlook*; 2017.

3. Davidson JW. The geology and prospectivity of Buton island, SE Sulawesi, Indonesia. In: *Twentieth Annual Convention of the Indonesian Petroleum Association*. Jkarta, Indonesia; 1991.

4. Masliyah J, Zhou ZJ, Xu ZH, Czarnecki J, Hamza H. Understanding water-based bitumen extraction from athabasca oil sands. *Can. J. Chem. Eng.* 2004;82(4):628-654.

576298 Chemical Kinetic Modeling Study of Heavy Hydrocarbons in Naturally Fractured Reservoirs through in-Situ Upgrading Technology

In Situ Upgrading Technology is new thermal-catalytic oil recovery technique that uses hydrogen for improving chemical properties of heavy oils contained in reservoirs, thus increasing recovery efficiencies. A chemical kinetic study was carried out to test hydroprocessing reactions within the matrix and fractures of a carbonate core pack at different temperatures and residence times.

The key elements involved in the present hydro-catalytic oil recovery process are ultra-dispersed catalytic particles, a hot fluid made of vacuum residue as a nano-catalyst carrier and low flow of hydrogen reactant. The study was conducted in a continuous mode, carried out in a laboratory unit that uses a carbonate core pack elementally representing a Mexican oil reservoir. The core pack set up was built up having pressure, temperature, rock characteristics and oil in place similar to real Mexican reservoirs. Cores were saturated with crude oil at 100 ^{o}C and 1100 psig; then a hot fluid with ultra-dispersed catalytic particles was percolated at 350 ^{o}C and 1450 psig to irreversibly deposit the catalyst particles over the rock. After a critical nano-catalyst deposition was built inside the core pack, a hot fluid without catalyst was flowed through the core under varying conditions of residence time and temperature, keeping the pressure constant (1450 psig). Collected products were characterized using Gas chromatography Simulated Distillation, P-Value stability test, Micro Carbon Residue and, Sulfur and Nitrogen elemental contents to assess the extent of hydroprocessing.

Experimental results showed stable and high quality recovered oils. The P-Value of the most severe condition was 1.95, comparable with the original value of the crude oil in place (2.0). A low tendency of the studied Mexican oil to produce coke was observed during the In Situ Upgrading test by passing many pore volumes of the VR with a micro-carbon of 24.24% through the core-pack. The oil coming out of the core-pack showed a micro-carbon residue of 17.15 % wt. under the most severe upgrading conditions (350°C, 1450 psig, and 72 h residence time). This oil resulting from VR shows no much difference compared to 16 %wt. measured for the original oil in place.

The present work provides insights into the oil reactions occurring in the matrix and fractures of the rock during the process In Situ Upgrading, providing important clues about interactions between the rock, the hot fluids contained in the reservoir and hydrogen reagent.

552507 Switchable-Hydrophilicity Triethylamine:Formation and Synergistic Effects of Asphaltenes in Stabilizing Emulsions Droplets

In this study, SHT (switchable-hydrophilicity triethylamine, $[Et_3NH]\cdot[HCO_3]$) has been synthesized and instrumentally characterized by Fourier transform–infrared spectroscopy (FTIR) and ^{13}C nuclear magnetic resonance (NMR). The operational synthesis conditions of SHT were optimized and determined at 25 °C, Et_3N/H_2O volume ratio of 1:2 and CO_2 injection rate at 300 mL/min.When it was used to extract heavy oil from unconventional oil ore, it was found that it could break maltenes-in-water emulsions. When asphaltenes were present in the oil phase, it was observed that SHT could cooperate with asphaltenes. These results indicated that SHT works with asphaltenes, leading to synergistic effects in stabilizing oil–water (o/w) emulsions.

566488 Nanoscale Static and Dynamic Behavior and Optimized Controlling Technology of Gas Channeling in Shale Oil Reservoirs

Nanoscale Static and Dynamic Behavior and Optimized Controlling Technology of Gas Channeling in Shale Oil Reservoirs

Kaiqiang Zhang, Songyan Li, Na Jia

Keywords: Phase behavior; Flow mechanics; Gas channeling; Optimized controlling technology; Shale oil reservoirs.

OBJECTIVES/SCOPE:

In this paper, static phase behavior and dynamic flow mechanics of gas channelings in shale reservoirs during CO_2 injection processes are experimentally and theoretically studied, on the basis of which two controlling technologies are proposed and evaluated. The developed gas channeling controlling technologies are finally applied in an oilfield located in Northeast China, whose implementation schemes, including the technical parameters and surface/subsurface infrastructures, are optimized.

METHODS, PROCEDURES, PROCESS

In theory, a nanoscale-extended equation of state (EOS), which considers strong confinement effects and intermolecular interactions, is developed to calculate the vapour-liquid equilibrium of CO_2 streams in nanopores. Moreover, complex mathematical models of wellbore pressure distributions and heat transfer are proposed.

In experiment, first, a series pressure-volume-temperature (PVT) tests of the crude oil-CO_2 system are conducted at P = 5-20 MPa and T_{res} = 97 °C. Second, fifteen laboratory dual-permeability CO_2 coreflood tests at five injection rates and reservoir conditions are completed to investigate the characteristics of gas channelings. Finally, some additional PVT and coreflood tests are performed to develop and evaluate the controlling technology in shale oil reservoirs.

RESULTS, OBSERVATIONS, CONCLUSIONS

The nanoscale-extended EOS is accurate for calculating the phase behavior of CO_2 streams in bulk phase and nanopores. It is found that the phase behavior in nanopores become significantly different, for example, CO_2 pressures in nanopores are always larger than the bulk pressure at the same volume. The shifts of critical properties are concluded to dominate the phase changes from bulk phase to nanopores. Phase transitions from subcritical to supercritical CO_2 are critically important, which is determined to occur at the respective depths of 1052 and 795m for Wells DB33-2-2 and DB33-2-4 in the target reservoir. The gas channeling, which is detrimental to recover oils, easily occurs at high gas injection rates and/or in large permeability (fractured) cores. In this study, the oil recovery factor is found to increase as high as 20-40% before gas breakthrough/channeling. On the basis of results from this study, the HY-2 CO_2 foam, which owns a stable foaming concentration of 1% and resistance factor of 40-80, and polymer gel of AP-P4(0.3%)+YG107(0.3%), whose optimum conditions are at pH range of 3-8 and temperatures of 40-97°C, are determined to effectively control the gas channeling. The comprehensive implementation scheme for controlling gas channelings in shale formations is optimized. Also, a foam surface generator and a subsurface injector with the temperature and pressure tolerances of 200 °C and 50 MPa are designed and applied in the oilfield.

NOVEL/ADDITIVE INFORMATION

Static phase behavior and dynamic flow mechanics of CO_2 gas channelings in nanopores are studied for the first time. Accordingly, the optimized technology and implementation schemes are initially proposed and successfully applied to control the gas channeling and enhance oil recovery in the actual shale oil reservoir.

568582 Development of a Hydrogen Impurity Enrichment Device Using Pd Alloy Membranes to Support the Growing Hydrogen Economy

Due to the damaging environmental effects of fossil fuels from the transport industry, national and international targets have been set in order to reduce CO_2 emissions. One solution to this problem is the further adoption of fuel cell vehicles which use hydrogen as a fuel source. In the UK it is expected that 1.6 million fuel cell vehicles will be on the road by 2030, with 1,100 hydrogen refueling stations in operation to support this growing sector.

A major problem with fuel cell vehicles is that these systems are extremely susceptible to degradation by impurities. ISO 14687 – 2:2012 specifies the maximum impurity levels of 13 impurities that can be present in fuel cell hydrogen. There is a large analytical challenge associated with validation of the standard, as most impurities are difficult to measure with standard techniques.

A Hydrogen Impurity Enrichment Device has been developed at NPL. Using a Pd-Cu coated Pd-Ag membrane ppm levels of impurities within a hydrogen sample are concentrated to levels where they can be analysed using low-cost equipment. A novel tracer enrichment method was developed which reduced the uncertainty of the measurement by ~60%. It was discovered that a key problem with the accurate of the device was the adsorption or reaction of impurities within the enrichment vessel.

Current work involves optimization of the membrane used and device operating conditions. Tests are underway using binary and ternary Pd alloys manufactured through electroless plating and magnetron sputtering in order to maximise impurity resistance and lower the cost of device, and maximise the enrichment factor.

569501 Shock Tube Study of Co-Optima Biofuels Combustion

Global consumption of fossil fuels and the production of greenhouse gases continues to rise, causing deleterious effects to the environment and all its inhabitants. The transportation sector is one of the largest culprits in greenhouse gas emissions— in 2017, 29% of all energy consumed in the U.S. was by the transportation industry and 89% of that energy came from petroleum based fuels [1]. In order to curb these increasing levels of greenhouse gases while still meeting the world's energy demands, alternative fuels are essential. The Co-Optimization of Biofuels and Engines (Co-Optima) has sought such alternatives in biofuels, because they are readily produced in large quantities through renewable means, have the potential to decrease greenhouse gases compared to petroleum based fuels, and may even provide similar energy outputs to gasoline. Co-Optima have identified several biofuels of various chemical classes, which show potential to be used in future engines to increase fuel economy and efficiency while decreasing greenhouse gases. However, before any of the identified fuels can be consumed by the general public, well-validated chemical kinetic mechanisms must be constructed so that the complete reactivity and behavior of these fuels is understood.

Validating kinetic mechanisms requires experimental data which is easily obtained in a shock tube—a device that simulates engine relevant conditions via gas compression from shock waves. Using this device, ignition delay time and species time-histories can be obtained during the fuel decomposition under engine relevant conditions and compared to the outputs of these kinetic mechanisms. Valuable molecules for time-history comparisons include carbon monoxide, ethylene, and formaldehyde. Carbon monoxide and ethylene are important because in hydrocarbon decomposition, these are two of the most abundant molecules formed and in oxidation processes carbon monoxide is the direct pathway for carbon dioxide formation, the major contributor to elevating global temperatures. Formaldehyde is also an important intermediate because it is an essential pathway for many of the major species formed through hydrocarbon decomposition. Therefore, in this work the Co-Optima fuels 2,4,4-trimethyl-1-pentene, prenol, isoprenol, cyclopentanone, and methyl propyl ether are investigated behind reflected shockwaves to obtain these key parameters for mechanism validation.

The conditions and data collected for each fuel varies depending on the validation targets needed for each specific fuel. 2,4,4-trimethyl-1-pentene is investigated between 1186-1414 K at 9.5 atm and an equivalence ratio of 1.0, ignition delay times and carbon monoxide time-histories are reported. Prenol and isoprenol are studied between 1269-1472 K at 9.4 atm and an equivalence ratio of 1.0. Cyclopentanone oxidation and pyrolysis have been examined, carbon monoxide time-histories and ignition delay times are collected during the oxidation of cyclopentanone and conditions span a temperature range of 1165-1327 K at pressures of 1, 6, and 8.5 atm. For cyclopentanone pyrolysis, the fuel depletion and formation of carbon monoxide and ethylene are measured at conditions of 9.5 atm and 1079-1393 K. The oxidation and pyrolysis of methyl propyl ether (MPE) has also been investigated. Formaldehyde time-histories were measured during the pyrolysis of MPE and spanned conditions of 1180-1474 K at pressure near 2.5 atm. During MPE oxidation, carbon monoxide time-histories and ignition delay times were recorded at conditions of 1313-1511 K and 9.5 atm. Data for all species are compared to their respective kinetic mechanisms previous available in the literature or preliminary mechanisms that are being developed for the Co-Optima program. Discrepancies between the experimental and theoretical data are highlighted, and where appropriate, suggestions are offered for changes in kinetic rates.

1-Administration, U.S.E.I., 2017. https://www.eia.gov/energyexplained/?page=us_energy_transportation

574074 Production and Characterization of Biorenewable Thermal Deoxygenation Oil/ Ultra Low-Sulfur Diesel Blends

Thermal DeOxygenation (TDO) oil is a type of highly deoxygenated biocrude produced through non-catalytic thermochemical conversion of levulinate and formate salt mixtures. Properties of TDO oil such as low oxygen content (< 5 wt %), broad boiling point range (75-550 °C), and high aromaticity suggest that it is a suitable candidate for making fuel blends with petroleum products. In this study, TDO oil and upgraded TDO oil were blended with Ultra Low-Sulfur Diesel (ULSD) and analyzed for chemical and physicochemical properties. The TDO oil was produced using a mixture of calcium levulinate and formate salt mixture at 450 °C in a 50 L semibatch reactor and hydrogenated using Ni/SiO_2-Al_2O_3 (65 wt % Ni) catalyst in a continuous flow tubular reactor. The crude TDO oil and hydrogenated TDO (HDO) oil were distilled to produce 200- 325 °C diesel range distillate. Sixteen different fuel blends were prepared by blending four different blend stocks, namely, crude TDO oil, HDO oil, distilled TDO oil, and distilled HDO oil with ULSD in four different volume percent (5 %, 10 %, 15%, and 20%). The fuel blends blend stocks were analyzed for specific gravity/ API gravity, viscosity, heat of combustion, hydrogen content, cetane number, boiling range distribution, lubricity, carbon residue, acid and base number, and hydrocarbon types in middle distillate according to American Society for Testing and Materials (ASTM) methods.

Some of the properties such as specific gravity, viscosity, hydrogen content, and net heat of combustion change linearly with the blend percent. Specific gravity and viscosity of all the fuel blends exceed or fall within the specified range of ASTM D975 No.2 distillates and MIL-DTL-16884M fuel specifications. Hydrogen (H) content (wt %) of TDO oil and TDO oil distillate fuel blends are comparatively lower than that of HDO oil and HDO oil distillate fuel blends. However, H content of all the fuel blends except the 15 % and 20 % blends of TDO and TDO distillate oil meet the MIL-DTL-16884M specifications. Net heat of combustion of all the fuel blends are approximately 42 MJ/kg or above. Overall chemical and physicochemical properties of HDO oil fuel blends and HDO oil distillate fuel blends are closer to the properties of ULSD.

561915 Acid-Catalyzed Rearrangement of Tetrahydrotricyclopentadiene: Synthesis of High-Density Alkyl-Diamondoid Fuel

Since aerospace vehicles like missiles and spacecrafts are always volume-limited, often operated in cold climates and at high altitudes, some features of the fuels such as high density, large volumetric energy content, low freezing point, and low viscosity become considerable important. In the development of fuels for aerospace vehicles, synthesis of alkyl-diamondoid fuels has fascinated scientists for many years. Here we reported a facile route to synthesize alkyl-diamondoids (methyl-1,2-tetramethyleneadamantane is the primary product) *via* solvent-free rearrangement of tetrahydrotricyclopentadiene simply catalyzed by acid. Different from previous ionic liquid catalysis with tremendous dosage, acid catalysis (simply use $AlCl_3$) makes the synthesis of alkyl-diamondoid fuels more facile and feasible. The two-step catalysis and solvent-free condition are favorable to the reaction towards alkyl-diamondoids, and the supreme content of alkyl-diamondoids in the liquid phase can reach 85.6% with carbon yield of alkyl-diamondoids of 60.5% catalyzed by low dosage of catalyst (10.7wt%). The reaction kinetics of rearrangement has been investigated, and the reaction rate constants and apparent activation energies were calculated based on the experimental kinetic data. And the rearrangement pathway of *endo*-THTCPD to alkyl-diamondoids is easier to take place while *exo*-THTCPD to alkyl-diamondoids is harder to occur. The resultant diamondoid-based product shows excellent low-temperature performance (freezing point less than -70°C) and a higher energy content (0.976 g/mL, 40.6 MJ/L) than reported diamondoids and widely used JP-10 fuel, thus is superior as a high-energy-density fuel.

577243 Advanced Gamma-Ray Densitometry (GRD) Technique to Measure Gas and Liquid Holdups in Bench Scale Hydrotreater Reactors

Advanced Gamma-ray Densitometry (GRD) Technique to Measure Gas and Liquid Holdups in Bench Scale Hydrotreater Reactors.

Mohammed Al-Ani[a], Hamza AlBazzaz[b], Muthanna H. Al-Dahhan[ac]

a Chemical and Biochemical Engineering Department, Missouri University of Science and Technology, Rolla MO 65409, USA

b Kuwait Institute for Science Research, P.O. Box 2488,13109 Kuwait

c Mining and Nuclear Engineering Department, Missouri University of Science and Technology, Rolla, MO 65409, USA

Email: aldahhanm@mat.edu

Trickle bed reactors are one of important hydrotreater reactors. Trickle bed reactors (TBRs) are multiphase catalytic reactors in which gas and liquid flow in concurrent downflow through solid particles (bed). TBRs have been used widely in the refineries field particularly in the hydrotreating processes (hydrodesulfurization, hydrogenation…etc.). The hydrotreating processes represent one of the essential processes in the industry due to contribute to reducing contaminant gases emission to the atmosphere. Bench-scale reactors broadly implemented to investigate the kinetics reaction and hydrodynamics as a part of scale-up approach or develop and improve industrial-scale system that called scale down approach. However, high importance for scale up and scale down approaches on the reactor performance, using advanced measurement techniques still in need to assess and measure different critical parameters with high accuracy in the bench-scale reactors.

Phase holdup in the packed bed reactors represents phase volume fraction in the bed. Gas and liquid holdup profiles are considered as a critical parameter due to significantly impact different design parameters. Gas and liquid holdup depend on gas flow rate, liquid flow rate, bed characteristics (shape and size of packed bed), and physical properties for fluids flow in the bed.

High accuracy and reliability to measure gas and liquid holdups are necessary for the bench-scale reactors. Gamma-ray densitometry (GRD) is one of effective industrial technique due to several attributes high reliability, non-invasive measuring which no need to interrupt the operating in the system, simplicity and flexibility in collect data along bed height within a diametrical profile. GRD technique constructed and developed in Multiphase Flow Reactor Engineering and Application Laboratory (MFReal) at Missouri University of Science and Technology. The general concepts for GRD are based on the photon count measurement. The basic principle of GRD technique is the differences in the density of materials in the column which the number of photon counts that reach the detector will be changed. The source Cs-137 with initial activity of approximately 250 mCi. The source housed inside a sealed chamber of lead from all direction except one that was represented as a collimator for a beam of radiation. Developed methodology applied to determine phase holdup in the multiphase flow reactor.

Many researchers have been investigated the hydrodynamics in TBRs using invasive techniques or even global measurements techniques. The results have been indicated high reliability for GRD to measure phase holdup profiles in the packed beds. Moreover, the effects of gas flow rate and liquid flow rate, and bed characteristics (shape and size of catalyst particles) on the phase holdup profiles at different bed height level axially in the hydrotreater catalytic reactor.

562545 Extraction of Essentials from Petroleum Sludge: A Review

Petroleum sludge is a composite mixture containing diverse measure of waste oil, wastewater, sand and mineral matter. Nowadays, handling of such petroleum sludge has become burden on the petroleum industries. The handling of sludge, its treatment, extraction of essentials and the discarding methods are the key matter of concern. It is also a key source of environmental pollution, spaced out from this, the extraction of essentials from the petroleum sludge can be carried out in order to fulfill the fresh oil requirement and evade the wastage of useful materials in form of sludge.

Disposal of sludge is a massive substance of concern as it causes environmental troubles. Sludge when disposed openly on land can lead to soil pollution, also when disposed in water body causes water pollution and harm to marine life. Recycling is the most desirable environmental alternatives for handling oily sludge, due to the possibility of recovering valuable oil for reprocessing, reformulating, or energy recovery. A range of methods which can be used for the separation of essentials from the petroleum sludge are solvent extraction method, centrifugation treatment, surfactant enriched oil recovery method, sludge pyrolysis, microwave irradiation, electro-kinetic method, ultrasonic irradiation and froth floatation method.

There is requirement of proper separation treatment of petroleum sludge before the disposal should be done to the outer world. In the present work, all the various disposal methods including incineration, stabilization or solidification, oxidation treatment and bioremediation was analyzed. The techniques of disposal under bioremediation include land forming, bio pile or composition and bio slurry treatment. The review work leads to the selection of the separation techniques for sludge purification.

558693 Deployment of Yeast and Enzyme Technologies for Sustainable and Cost-Effective Ethanol Production

Ethanol is a biodegradable, high-octane motor fuel derived from sugars, starches and cellulosic materials found in plants. Ethanol is part of the solution to reducing fossil fuels dependency, lowering fuel prices, creating rural jobs, boosting farm economy and preventing climate change. DSM is a frontrunner in the development and deployment of yeast and enzyme technologies contributing to the above. On one hand, advanced yeasts are nowadays steadily conquering the corn ethanol market. This is triggered not only by the low margins but is also a logical next step in plant efficiency improvement. Advanced yeasts offer increased ethanol yields, mostly by reducing glycerol formation. Often, this higher yield performance comes at the expense of reduced robustness and/or increased by-product formation. DSM has developed and commercialized a disruptive yeast technology, called *e*BOOST™, for yield enhancement (up to 5% higher yield than conventional yeasts with up to 75% reduced glycerol production), free of any by-product formation, generating also additional output for the ethanol producers.

On the other hand, utilization of non-food sources, such as lignocellulosic biomass, is a sustainable option for producing cellulosic ethanol. Enzymatic hydrolysis of lignocellulosic feedstocks into fermentable sugars is one of the most complex processes applied on production scale, since it involves the combined effect of a cocktail of enzymes acting in a series of chain reactions on multiple substrates (glucan, xylan and arabinan). Process aspects related to the enzymatic hydrolysis are also very important on production scale. For instance, fast viscosity reduction results in improved heat and oxygen transfer, allowing processing at higher dry matters and realizing better asset utilization. DSM has also developed commercially competitive cellulolytic enzymes that are uniquely active at high temperature and low pH, offering many benefits and product differentiation.

The current paper will demonstrate a proven and stable track-record of the *e*BOOST™ yeast technology on commercial scale fuel ethanol production, delivering what it promises even under challenging process conditions (e.g. high temperature, low pH and high organic acids), creating added-value for the customers. Additionally, we will demonstrate that the DSM cellulolytic enzymes' distinctive application conditions highly help limiting the adverse effects of contamination, which causes ethanol yield losses, and thereby eroding profit margins.

576758 "Dock and Lock" Covalent Assembly of Designer Cellulosomes

Lignocellulosic bioethanol offers an alternative energy source to fossil fuels while reducing lignocellulosic waste. Regrettably, this process is not yet economical due to the recalcitrant nature of lignocellulosic biomass. Microbes have evolved mechanisms to efficiently degrade lignocellulosic biomass through large enzyme assemblies termed cellulosomes. In cellulosomes, enzyme assembly is mediated through the interaction between cohesin domains that make up the noncatalytic scaffoldin proteins and dockerin domains that are fused to enzymes. The use of orthogonal cohesin-dockerin interactions has further led to the development of designer cellulosomes in which the number, ratio, and position of protein components can be controlled. However, the natural cohesin-dockerin assembly is mediated through noncovalent, reversible interactions that inevitably result in the partial dissociation of designer cellulosome components, an issue that will be compounded as the size of designer cellulosomes increases.

In this study, we developed a "dock and lock" method to enable irreversible assembly of cellulosomes by introducing a covalent bond to the cohesin-dockerin interaction. Specifically, homology modeling, docking, and structural analysis were performed to design cohesins and dockerins with single cysteine mutations so that cohesin-dockerin guided assembly causes the formation of a disulfide bond between the two proteins. Our data showed that introducing the cysteine mutations did not affect the cohesin-dockerin interaction, but created a covalent complex that has a greatly reduced dissociation rate and much improved stability without dissociation even under boiling conditions. As a result, these engineered cohesin-dockerin pairs improved the assembly efficiency of designer cellulosomes, which is further translated to enhanced catalytic properties. The method reported here provides a simple way to form covalent designer cellulosomes and can be seamlessly incorporated into any large protein assemblies with minimal design implications.

578763 Bioethanol Production Using Simultaneous Saccharification and Fermentation (SSF) Technique with Dioscorea Rotundata (Yam) As Raw Material

The low water requirements of tubers as cassava and Dioscorea (yam) and their high starch mass concentration are an opportunity to use them on a market that can enhance the agro-industry of the on development countries. The scope of the research includes a decrease in the energetic requirements and the possibility to explore new crops as raw materials for producing bioethanol.

This research evaluates the productivity and yield of bioethanol production, using Simultaneous Saccharification-Fermentation (SSF) technique with Dioscorea Rotundata (yam) as raw material. The pretreatment began with a starch gelatinization for 30 minutes at 68°C; then starch was hydrolized with amylase for 60 minutes over 90°C and glucoamylase for 30 minutes at the same conditions. After the hydrolysis beginning, the system was cool to 30°C and inoculated with yeast (SafCider-Fermentis). The samples rested on controlled warming for 26 hours, taking samples from them. Simultaneously, the technique applied to measure the reducing sugars concentration was DNS and ethanol concentration was calculated by HPLC.

The yam concentration used on each sample were 10 w/w%, 12.5 w/w%, 15w/w% and 18w/w%. However, the results show that the 10 w/w% sample achieved the best result, in contrast to the suggested concentration of 12.5% w/w% (J. Murgas & M. Vasquez, 2012), who last 50 hours on fermentation with Separated Hydrolysis-Fermentation (SHF). In addition, the fermentation occurs faster than the references that have employed SSF method in other types of crops, with approximately 22 hours of fermentation and at least 33% of the reduction in fermentation time. Furthermore, the process achieved a 90% of enzymatic hydrolysis yield, getting the 67 w/w% of total yam mass added as reducing sugars. On the other hand, the best ethanol yield obtained was 46.67 l/ton and 0.188 g/l h as productivity. Nevertheless, fermentation yield was 12.9%, a low value if it is compared with the typical fermentation yield that is between 49.5% and 52%. This deviation was related to a low ethanol selectivity of the yeast used on the samples.

576473 Prediction of Yields and Composition of Products from Torrefaction

Ireland's 2015 Energy White Paper proposes the guidelines for a radical transformation in the energy sector in Ireland and a large reduction of CO_2 emissions from 2015 to 2050. Torrefaction is a promising pre-treatment technology, which generates the value-added material from solid waste. In order to improve the properties of briquettes, it is necessary to understand how primary char properties are correlated with feedstock composition and operating conditions since char is formed as a result of the reactions between carbonaceous feedstock matrix, volatiles and inorganic matter. In this study, olive stones were torrefied at various temperatures and holding times in a lab-scale fixed-bed reactor and at one fixed temperature and residence time in a continuous industrial torrefaction unit. The aim of this work is to obtain knowledge about effect of different lignocellulosic compounds and operating conditions on the composition and yields of torrefied biomass. In this study, the kinetic model was validated against the own experimental results. The results of the present work indicated that the catalytic effect of alkali metals on the product yields and composition is significantly less during torrefaction than at temperatures greater than 300°C. The conversion and product distribution of volatiles during torrefaction were related to the hemicellulose transformation. The potential key role of hemicellulose and lignin on the composition and yield of torrefied biomass is highlighted, the issue of torrefaction temperature is discussed. Results showed that the proposed kinetic model for the torrefaction and gas phase reactions is relatively simple and predicts the product yields and composition accurately.

AUTHOR INDEX

AUTHOR INDEX

AUTHOR INDEX

AIChE
120 Wall Street, FL 23
New York, NY 10005-4020

ISBN 978-1-7138-0538-0

9 781713 805380

CURRAN ASSOCIATES INC.
proceedings.com